国外大师谈话录

圣地亚哥·卡拉特拉瓦与学生的对话

国外大师谈话录
圣地亚哥·卡拉特拉瓦与学生的对话

[美]麻省理工学院 编

张育南 译

中国建筑工业出版社

著作权合同登记图字：01-2002-3941号

图书在版编目（CIP）数据

圣地亚哥·卡拉特拉瓦与学生的对话/（美）麻省理工学院编；张育南译.
—北京：中国建筑工业出版社，2003
（国外大师谈话录）
ISBN 7-112-05918-6

Ⅰ.圣… Ⅱ.①麻…②张… Ⅲ.建筑设计-理论 Ⅳ.TU201

中国版本图书馆 CIP 数据核字（2003）第 055415 号

Copyright © 2002 Massachusetts Institute of Technology
Chinese translation copyright © 2003 by China Architecture & Building Press
First Published in the United States by Princeton Architectural Press
All rights reserved

本书经美国普林斯顿建筑出版社和美国麻省理工学院正式授权我社在中国翻译、出版、发行本书中文版

策　　划：张惠珍　黄居正
责任编辑：戚琳琳　马鸿杰
版式设计：彭路路
责任校对：王金珠

国外大师谈话录

圣地亚哥·卡拉特拉瓦与学生的对话

［美］麻省理工学院　编
　　　　　张育南　译

中国建筑工业出版社出版、发行（北京西郊百万庄）
新　华　书　店　经　销
制版：北京嘉泰利德制版公司
印刷：煤炭工业出版社印刷厂

开本：850×1168毫米　1/32　印张：5　字数：130千字
2003年9月第一版　2003年9月第一次印刷
定价：**20.00**元
ISBN 7-112-05918-6
　TU·5196（11557）

版权所有　翻印必究
如有印装质量问题，可寄本社退换
（邮政编码 100037）
本社网址：http://www.china-abp.com.cn
网上书店：http://www.china-building.com.cn

目　录

序　6

拉菲尔·巴拉斯与斯坦福·安德森

前言　9

材料与建造过程　10

力与形式　23

运动与形式　33

结语　43

序

1995年的11月,当我漫步于西班牙的巴伦西亚时,为一座形体简洁但非同寻常的美丽桥梁所折服,我向人打听设计它的建筑师,立即知道了圣地亚哥·卡拉特拉瓦的名字。我必须承认,当时我并不了解他。但是通过接触,我对他迅速了解并熟悉起来。在我的沟通下,我的一位同僚赫伯特·爱因斯坦教授,代表麻省理工学院对其发出邀请,卡拉特拉瓦欣然接受邀请前来进行学术访问。这次访问使我们得以会面并举办了一系列的讲座。本书与其相关网站(http://web.mit.edu/civenv/Calatrava/),都是从卡拉特拉瓦和专家与学生们在1997年的那三天访问期间的发言中整理的。

听了卡拉特拉瓦的讲座,我更加明晰自己为什么要选择土木工程专业,更激发了我的创作欲望,去解决功能与美学之间的矛盾,创造为世人所留恋的作品。我猜想,所有的工程师都会像孩子一样怀着纯真的梦想去实现自己的理想,然而非常不幸,我们的教育体制压制了这种想像力的发挥,建筑学与工程学的结合成为泡影。创造性被程式化和单一的专业训练所湮没。

卡拉特拉瓦向我们展示了建筑与工程相结合的设计师的素质,他所设计的建筑和桥梁反映了对工程的深入理解,例如一个经典的拱,它的结构总是与受力情况相吻合,反之亦然。力学的要求融入结构形态之中,没有多余的构件,设计的目的性实在而明确,作品浑然天成,宛自天开。

实际上卡拉特拉瓦经常从人类的骨骼构造中获得灵感,因为那是自然赐予人类最具美学和功能的物体。他所设计的建筑如同人体一样,用简单而独立的构件组合成庞大而复杂的系统。

并非所有人都能像卡拉特拉瓦一样成功,很少有人能够具有他那种天赋和对艺术的感悟力,但是每个从事工程的设计者都有条件通过自己的努力使项目更具创造力,每位建筑设计者也可以在了解结构与机械的特性之后更富于想像力。如果我们照此努力,那么我们的专业会远比现在更具吸引力。

拉菲尔·巴拉斯
Bacardi and Stockholm Water 基金教授
曾任麻省理工学院土木与环境工程系教授

序

工程设计与建筑设计的分离由来已久，现在，至少在美国这已经是放眼皆准的事实。这种技术与艺术的分离使得两个方面都蒙受损失。从建筑角度来讲，失去了完美设计所应具有的勃勃雄心；而从工程角度来看，设计成为一种程式化的操作，失去了对社会、环境和美学标准上的对话。

作为一所高等学府，麻省理工学院当然离不开我们当前所处的时代与环境，但我们可以放心大胆地想像工程与建筑更深入融合的情景。值得庆幸的是，总会有一些人怀着极大的梦想创造出惊世之作。在建筑界，人们都推崇伦佐·皮亚诺和他的"建筑工作室"，那里的优秀创作是以其对建筑在各个领域内的大胆突破而闻名的。但是对于皮亚诺而言，同许多优秀的设计师一样，他和他的工作室在一种精益求精的合作过程中体现出难得的集体创作的力量。当人们想起伦敦的阿鲁普公司和巴斯的哈泼尔德公司、还有巴黎的 rfr 等设计工作室时，所有的创作集体都超越了单独每一位设计合伙人的水准。

特别是在桥梁、基础设施和大跨度建筑物的设计当中，人们发现最终成功把握方案的设计师必须在技术与美学领域都有突破。我们通过举办菲里克斯·坎德拉系列讲座*，终于接触到了许多位于这种领域的先锋人物，其中包括：海因茨·伊斯勒、minoru kawaguchi、克里斯汀·梅恩和约尔格·施莱克。上述的每一位大师都在他们的设计中遵从了非常严谨的科学规律。然而仍然可以在这些工程中体现他们的个性色彩。

圣地亚哥·卡拉特拉瓦作为一名建筑与工程相结合的设计者，坚定不移地追求科技与艺术融合的道路，他对于自然界形态的发现（尤其是人体的骨骼）、他埋头苦干地工作和他本人的天赋，共同体现于其对形体、空间、光学乃至运动的把握。他神奇的工程经验不仅使自己的构思得以实现，而且完成了个人独创与科学规律之间大胆的开创性对话。

卡拉特拉瓦在讲座中，用富于活力的表情和极富表现力的图画来描述自己完美的设计过程。我们希望这种动态的过程同样在记录的文字中有所体现，对促进工程技术学与建筑学的结合有所裨益。

斯坦福·安德森
麻省理工学院建筑系建筑与建筑历史专业教授

---

* 菲里克斯·坎德拉系列讲座于 1994 年由美国纽约结构工程学会、现代艺术博物馆、普林斯顿大学建筑系以及麻省理工学院共同承办。除了宣传和表彰坎德拉本人在建筑与工程领域的成就之外，该系列讲座力图通过宣传当代技艺结合的工程典范，以促进这些领域的学术进一步发展。

前　言

女士们先生们，非常庆幸有机会在这里演讲。我曾经在巴伦西亚和苏黎世经历了14年漫长的学习生涯，之后又开始了自己的建筑设计与结构设计生涯，如今16年又已过去。我一直孜孜以求紧张地进行探索，能与像麻省理工学院这样的高等学府进行交流的惟一方式就是举办个人讲座，这也是我第一次结合自己的实践进行系列讲座。我认为时间非常恰当，因为这16年是我个人设计生涯中最重要的阶段，它标志着一代人的努力。而现在我的谈话对象是下一代的年轻人——那些会根据我的经验发现和寻找他们自己道路的年轻人，就如同我当年沿着前人的足迹寻找自己的方向一样。

我认为最好结合自己的工程经历总结这些经验，因为我所能做的也仅仅如此。我将回顾自己的工作，尽力向诸位介绍这些年来我思想形成的主要历程，还有从一件作品到另一件作品的一次次实践中思想逐步深化的过程。

材料与建造过程

我首先提到的是材料,因为对我而言,材料是最基本的建筑要素。最终,人们在毁坏的建筑废墟中发现的只是石头。所以在我看来,研究建筑学的材料方面——它们承担着建筑学的物质层面——是非常重要和根本的。

我认为认识建筑学的第一步是去了解什么是混凝土、钢材、木头等材料以及如何应用它们,还有它们究竟表现了什么。它们最终可以构成什么样的造型,这就是今天我竭力向大家展示的,这将从我最初的作品一直贯穿到最近的作品。

在这个用玩具搭建的小雕塑造型中,旋吊石头的力贯穿物体的每一个组成部分,通过石材、木材和绳索(缆线)甚至是钢材所组成的物品传递着。这看起来十分简单,但实际上却颇有学问。这里所有的玩具都处于受拉状态,线团产生向外的倾覆力,但它本身处于受压状态。连这些物品的色彩也呈现一种简洁却精心的格调。毋庸置疑,这里充分表达和揭示了在空中悬吊石块的情形。

我在瑞士的阿拉干的沃连高级中学承担了一个项目,是在几栋现有的建筑之间进行增建的工程。我设计的项目包括一个入口、一个中心报告厅、图书馆的顶棚和大会议室的屋顶。在设计过程中,我几次根据材料的变化修改方案。有些部分运用混凝土和钢材的组合方案,有些则运用玻璃和钢材的组合方案,还有些部分运用了混凝土和木材的组合方案。依据对不同材料进行测试的结果,运用不同的材料特性解决不同的需求——在这里我要向大家提及另一个非常重要的话题——用光来控制各种不同空间的特性，那对于我来说是件非常有趣的事情。

入口的构思是从现存建筑的布局和规划中得到的，平面呈梯形，我的方案是切去了其中的一个对角，形成了由一个拱券连接的两个平面呈锥形顶盖。每个顶盖都偏向各自的方向。在它们的交接处有一个截面呈管状的拱，起到抵抗扭矩和排水天沟的作用。即便如此，管状物本身还是承受了很大的扭矩。我在这里用它来作为立面上的衔接和屋顶覆盖物，用以协调所有的建筑构件以形成建筑风格上的统一。

与此同时，每一个建筑构件又应清晰地表现，尤其是在立面上更为重要，在本方案中表现为树叶或手掌的形式，一种非常具有形式感的造型本身就是设计的组成部分。这当然是在我对项目有了初步的了解之后而产生的构思，这样做可以赋予建筑物以令人惊叹的自然形态。单纯对事物的观察如同建筑材料一样都激发着我的创作欲望。

沃连高级中学的第二个增建项目是一个入口空间。在那里，我设计了一个木制的圆形屋顶。屋顶的形象非常鲜明，其圆形被辐射状分割，分割后每一个部分呈V字型，使内部空间十分开敞。我用线型的、代表受压的构件取代了折角部分的构件。我将圆顶上不同类型的支撑构件各自分开，使人们领会到这些构件在视觉上共同构成了顶棚外围的一个圆环。这个承受拉力的环浮在半空，它是我与地心引力进行对抗的思想的具体体现，使圆环看起来非常明显又不含任何结构意义。屋顶真正的支撑——V字型构件的转角部分起到了拉结的作用。

光从圆形屋顶的后方进入，从它底部反射到三角形的折角部分。位于圆顶后面的部分在光的作用下仿佛消失了，所以使人感到整个屋顶好像飘浮在内部空间之上。显而易见，当你看到所有的构件组合在一起时，就仿佛与花瓣的形象突然有了些许默契。

在图书馆的设计中，对光的处理以及协调光与空间的关系变得十分重要。我认为自由墙体分隔和形成一束沿切线方向的光非常必要，就像路易斯·康或许曾经做过的那样。我希望让位于圆形中心的屋顶看起来像飘浮在空中一样，其主要的支撑是位于中央屋顶向之倾斜的柱子，它同时又是汇集屋顶雨水的通道。屋顶是由几片薄片组成的壳体结构，其折角设计得恰到好处，以确保屋顶不会向旁边移动，但整个屋顶的重量都坐落在柱子上，当光向下照耀时，就产生了沿着墙体呈切线的光束。

为了吻合图书馆储藏图书的功能，我构思了一本张开的书的形象。脑海中依然浮现着出现过几次的"屋顶浮在半空"的构思，这是为体现屋顶的轻盈而进行的处理。创造轻盈感是因为要与材料或静力学结构形成对比。在这里，通常表现屋顶感觉较重而承重的中柱却显得很轻盈，两种材料的对照与侧面采光的作用使整个屋顶仿佛飞了起来。壳体的基座不是普通的双曲面或抛物面，它是一本打开的书的形象。同时，壳体还仿佛是一只展翅欲飞的鸟。这是一种设计概念的叠合，你还可以从中柱托起的壳体中领略到叶子的形象。

至于大厅或者演讲厅的设计，我想在这里实现的构思非常简单——那就是通过暴露被抛物线拱所支撑的屋顶来体现墙体与屋面在结构上的分离。在壳体的两侧设计了天沟和采光的水平天窗，它们将自然光引入室内，非常微妙地表现了壳体的结构特征和那些丰富细腻的结构构件，使这些将屋顶的荷载传递给拱的构件感觉非常通透。这些斜撑都是由3英寸（约7.6厘米）见方的木材构成，抛物线拱和上面支撑屋顶的其他拱均由复合木材加工而成。这个空间感觉非常亲切，其室内构件的风格和光影效果也有助于这种亲近感的形成。

伫立在入口处的屋顶覆盖物，也同样体现了棕榈树一样的植物形象，它们处理得也许有些雕塑感过于强烈——我试图在此表现荷载从抛物线拱传递到柱子上的过程。许多人将柱子的顶端看成是简化了的爱奥尼柱头，实际上并非如此，它更像牛腿的顶端。

柱子是由混凝土预制而成的，我原本非常喜欢素混凝土——即在施工场地上现浇出的混凝土，但只要你把握它的肌理和造型特点之后，预制的混凝土也会产生非常强烈的表现力。在我的巴伦西亚母语中称这种混凝土为"formigó"，它源于单词"forma"；在西班牙语中，用来表达混凝土的单词是"hormigón"，以字母

"h"替代了巴伦西亚语"formigó"的字母"f"。"formigó"的本意是赋予材质以形状,它非常恰当地表现了混凝土的特征,用预制混凝土构件,还可以使你在选择材料的形状、纹理和其他特征上有非常大的自由度。

大厅的柱子是用非常节省的方式预制成的混凝土柱子,预制时将整根柱子分为两截,分别浇灌成型后再拼接到一起。这种方法还有一个优点,那就是浇灌时形成的表面就作为柱子的最终面层;在现场你看不到任何混凝土施工的工地。

"德国科斯菲的厄恩斯廷工厂仓库"同样是我早期的作品,它是如何将过去的仓库进行改造并使之具有新的含义的典型实例。我所要做的第一步是选择材料,它们必须非常便宜,所以我决定选择混凝土和生铝板,即你到处都可以见到的、感觉粗犷的标准材料。你可以买到呈卷帘形的或平铺的铝板,你可以用多种方式加工它,以使它们非常易于应用。

我们来探讨如何使这栋建筑得以改造——不仅是在材料上,而且是在理念上对其进行改造。用材的限定在很大程度上形成了制约。在这里我们的构思围绕着一个主题——如"拼图"般的效果,它使得我们可以在某种程度上相对自由。在这里所讲的"拼图",意味着每一个立面都会因两种材质的拼接而形成不同的图案。每个立面的处理手法可以有非常大的差距,只须运用材质来统一它们。

南立面采用弯曲的铝板饰面，将它们加工为呈正弦曲线的形状。由于曲线凸出的部分依靠内侧的横向构件连接，因而在立面上形成了双重的刚度。曲线提供了一个方向上的刚度，而弯曲的铝板本身还有另一个方向上的刚度，所以它们允许我们建造很高尺度的立面。

为了强调曲面的动感，分别在立面的顶部和底部斜切了一下，使得从前方看去，立面上的正弦曲面更为突出。从下面仰视整个建筑，曲线的形状清晰而明了。委托商告诉我，因为安全照明的需要，外部需要许多灯光照明，因此我在立面上设计了灯柱。灯柱的照明强化了材料表面的特征，在设计方案中使灯柱探出建筑立面之外，从而使建筑墙体上的阴影随着曲线而产生变化，同时在立面上形成了差异，构成非常富于艺术感染力的画面。

上面讲的是南立面，它的特点在于它的效果将随着一天中太阳的不同位置而发生变化。在曲面里你可以看到这里不仅隐藏着阴影，还点缀着太阳所形成的高光。弯曲的铝板反射阳光时会形成垂直的振荡。立面变化非常微妙，根据日光水平或垂直的入射角度，在一天中的每一刻呈现不同的效果。底部的斜切面是仿佛从混凝土的结构中飘浮而出的样子。

北立面，情形又有所不同，在那里只受天顶光的作用，即一种散射的环境光。那么如何在这种光照条件下形成匠心独运的立面效果呢？我在立面上采用了相当长的一段水平布置的"S"形铝板，尽可能地增大它的长度，这意味着 10 米或 30 英尺的距离。在整片板材的截面上位于中心对角线的部分，能够形成反射，在两重板材叠合的上部会产生阴影。然后同样，底部反光也较弱并被阴影隔断。这形成了一种在天顶光的作用下非常醒目的水平结构。

东立面，我使用了 17 米高的弯曲铝板，我们需要解决的难题是如何将其拼装成一体，并在顶部或底部用细小的构件将其固定。在初生太阳水平的照射之下，东面的场地非常平坦，那里还隐藏着一个外饰金属面材的升降机。为了将之与其所属立面区分，我们借助于比例——像很大的、铝制的鱼一样的比例，它们在折角至顶部可以折叠。

于是立面又一次被看成分离的图案，也就是我曾经所说的"拼图"，它就如同画卷一般展现在我们面前。但是你怎样才能将这些不同的画面，也就是立面连接在一起呢？从一开始，这栋建筑就好像天外来物，我们必须使它富于生机。西立面与东立面几乎完全一样，只是日落与日出的差别。这里同样用大面积的铝板解决了立面的形式，它们与三扇每日通行运货的卡车的大门结合在一起。这里有非常多的货车，许多停在门边等候。当大门开启时，货车开始穿梭，如果你在建筑或其他的尺度上瞭望大门时，建筑仿佛化作了一条巨鲸。当车流攒动时，就仿佛约拿的故事或圣 Exupery 的巨蛇和大象一样，进行着吞咽的场面。大家想想，这种设计上的游戏是多么重要啊!

在讲座开始时,我给大家看的雕塑是用玩具做成的。它是我利用家里给孩子们准备的玩具所作的构成练习——通过铅笔、挂窗帘线和其他我所能在家中找到的物品来进行的创作。在练习或构思中,最关键的是整体中任何一部分都成为不可或缺的。在一开始有些想法纯粹是自发和简单的。如何深入下去成为一个非常重要的问题。可是大家知道进行下去是一件很简单、很自然的事情。不积跬步,无以至千里。对于像鲸鱼的那些构思,在我看来不过是通过拼图和肌理赋予建筑以生机的方式。

在仓库门的模数设计上,沿袭了立面的模数。立面上一片片重复着同样的元素;仅仅是在长度上有所改变,这产生了一种非常美妙的构思,即可以使立面进行变形的构思。变形与变异的概念是形成建筑进化的必要条件——一个令人难以置信的必要条件。那不仅是因为通过变形的概念可以改变原有的机械和物质构件来适应新的建筑形式;而主要是因为我们对建筑的需求与对文化的思考仍处于一个不断成熟的过程,如今正是将这些新的内容融入建筑中的时刻。

你能想像几乎将一个完整的立面全部进行改造的情形吗?不局限于这个项目中的一半的立面的变化,而是整个立面都像窗帘一样从一侧拉到另一侧的效果。在我看来,这在今天是可行的,当然它对于新的一代的设计者和建筑的发展都是一个挑战。

苏黎世德斯塔德荷芬火车站，坐落在一处山坡地的边缘。早期，苏黎世湖曾经绵延至山脚——形成古冰河沉积形成的石堆。早期的居民在此填湖造田，形成一片平坦的土地并建造了房屋。后来，这里修建了铁路，两道铁轨将山地分成两半。分开后两边形成了非常不同的发展趋势——火车站至今仍然留存，在它的一侧是繁华的都市，而另一侧却被非常自然的植被所覆盖。

我们的项目是火车站的扩建工程。这是一个交通非常繁忙的车站，它在当地的铁路运输中发挥着非常重要的作用。我们的扩建必须考虑铁路本已将山地分为两段，而我们却不得不顺应它的趋势又将沿着山体将其弄得更深一些。最基本的构思是保留原来的边界，这样可以同时保护边界上的植被和景观。根据这些原则，我们设计了一片像墙一样永久地锚固在山体上的结构。这片墙支撑着上面的住宅，它们与用地的边界非常接近。沿墙建造了一些花园和花架，还保留着上部基地原有的风貌，使下面候车的乘客在等候火车的闲暇可以有机会到上面漫步。在铁轨下面，我们设计了带拱廊的商店，为的是使通向站台的道路更为安全和有趣。

还有许多自然的技术因素制约着建造。比如，我们工作的场地非常狭窄——最短处只有27米即90英尺宽，而且两面都有已建建筑。所以由于这种技术上的原因，我们不得不考虑发展地下空间，一直挖到地下14米也就是45英尺。同时，我们在建造时还要考虑全天候的铁路运行的因素。在建造过程中铁路运输从未间断，有时每隔2分钟就有一辆火车通过。在这样紧张的基地上，尤其是还要兼顾工期与安全的同时进行，改建对于我来说的确是一项挑战。

我当时非常紧张，因为这是我最初的兼任建筑师与结构工程负责人的几个项目之一。它虽然是一项大型公建，但也应该具有某种趣味性。所以除了考虑火车站本身的复杂功能和它与城市文脉的联系之外，我头一次尝试着运用了人体与解剖学的概念。我开始关注人的姿态，并从自己的手掌开始研究。张开的手掌，代表着真诚与开放，我从张开的手掌又想到了手掌的侧面，我选择拇指与食指之间的部分使之变为柱的形式，你可以在整个工程中看到这种形式被重复地频繁使用。

我还应该告诉大家，这个火车站是立交的而不是线平式的，火车通过一条地下隧道进站，再经过隧道出站。从剖面上看，隧道的两端向下弯曲，只有中间部分是平坦的，因此我们的车站就采用了微微拱起的拱结构。柱子微微倾斜以保持与拱顶相垂直，你不会发现它们之间的角度有什么变化，但如果它们垂直于地面，你或许会马上发现其倾斜角度的变化，而且它们在各个方向都是倾斜的。地下廊道的柱子也是类似的形式。车站的平面上同样呈弯曲状，这几乎纯粹是结构的需求所致。平面与轨道的剖面早已确定并由结构工程师设计好，但曲线却形成了一个280米长的拱券，差不多900英尺。由于在中间看不到拱的两端，所以它看起来比实际的长度更长。于是曲线的运用同时在平面与剖面上都促进了设计构思的发展。大家看，那旋转形体像圆环一样，强调了铁路的动势。旋转的几何体非常便于形成动感的建筑效果，尤其是通过重复的建筑元素。

你会非常有趣地发觉，实际上车站已不复存在，因为它已经消失于后面城市背景中的建筑屋顶之中。

我非常希望讲一小段用混凝土进行不同试验的插曲。我受"瑞士混凝土协会"的委托，为巴塞尔展览会设计了一个亭子。为了完成这个项目，我试图制造了一种使混凝土运动的机械。我们从前了解的不仅有轻质混凝土，还有普通混凝土，甚至所有种类的混凝土，但却很少有人知道运动的混凝土。不知大家是否知道弗雷西内工程师？他是试图用微粒混凝土先张法制造飞机机翼的发明者，或许这项技术根本就没有成功，但他却为此做出了各种努力。这说明在20世纪20、30年代的人们热衷于对这种材料进行探索。

在这座亭子的设计中，我想运用一种完全仿造人体的建筑形式。通过认识人体，我是说分析男人与女人的身体，我对形式的感性方面产生了兴致。于是，大家看到产生了像肋骨一样和类似于环形的构件。在这里应用混凝土是人们通常不会想到的一个非常巧妙而又折衷的解决方案。这不仅涉及运动混凝土的应用，还包括如何使它富于感官效果，比如类似肌肉与皮肤。这里的"有机"的想法是在探讨一片细部构件如何与另一片构件搭接的过程中逐渐产生的。细部的形体的作用，超出了精致的结构构件的范畴；它使我们可以如神话般地再现人的胸膛。

本章节中我所介绍的最后一个实例，是里昂的"沙特拉斯机场轻轨铁路车站"。提醒大家注意的是，这栋建筑全部是由非常细心而手艺高超的工匠们用手工制造的混凝土构件。大家也应留意在现场进行建造的过程中，将不同的构件组合在一起时的动人场面。与现场工作人员和睦相处是十分重要的，因为工程的关键在于匠人们的通力协助。实际上现场经常有人会有风险——他们当然不是建筑师，而是那些工作的可爱的匠人们。

车站由混凝土、铝材、钢材和玻璃等构成，还有一些花岗石的铺砌。用于建造它们的混凝土是白色的，由纯白的水泥制造，经自然搅拌后运至场地处，与浅灰色的沙子混合。这样就产生了一种外表非常轻盈的混凝土，它的颜色使你感到与现场十分协调。屋顶是铝板制成的，所以会反射阳光。站台的连廊以素混凝土浇筑而成，中间是大片的玻璃窗。步行道用花岗石和卵石铺砌，而其他硬地都以混凝土建造。敞开的巨大翼状屋顶，由钢材和玻璃共同制造。在这个设计里运用了非常多的材料，但其应用的法则却是经过精心推敲的。比如，钢材从来都不与地面直接相连——连接处的外表面都包上混凝土材料。起支撑作用的构件也从来不是直接架设在地面上的。

车站的翼形屋顶面向南方。在设计屋顶的倾斜角度时，我们考虑到太阳高度角的变化规律，使6月至11月间日光无法直接射入建筑内部，而其余时间太阳却可以温暖室内空间。这使我们在设计建筑时考虑适应季节变化的要求。

我非常希望能够根据人体的某一部分结构来创作建筑造型。许多这样的结构构件一起支撑着车站的敞廊。整个敞廊就是由这种元素组合而成的。

此外，整个车站和屋顶的构思最初源自于眼睛的形式。有人曾经说过如果拉菲尔失去了臂膀，他或许会成为一名优秀的建筑师，因为建筑师的作品是通过眼睛来完成的，这才是观察与判断和创造的源泉。视觉由两部分组成：两个长在面孔前面的眼睛；创造与联系事物的发现意识。

力与形式

如果我们将工程设计看作是一门艺术，然后我们还会想起过去曾经有一段时期，那时的建筑艺术与技术是统一的，如我前面所阐述的那样，那么我们可以想像在我们这一代，尤其是接下来的一代中，会出现艺术的重生。它不仅意味着我们从前人那里得到的，还有我们得到它们的前提条件——即通过自己的建筑实践对其所进行的诠释，以及一次次的重新认识。

在针对第二个主体进行演讲前，我还要向大家展示用我的孩子们的玩具制造的另一个小物品。它说明了如何将构件组合在一起，用于前面所讲的同样悬挂物体的目的。但在这个例子中，力的方向不是交叉而是彼此平行的。所以像游戏一样，通过它你能发现物体的重量可以解释许多事情，但却要受物体的受力法则的制约。

在我看来，桥梁的静力学结构是非常重要的。桥梁的静力学研究或许只是其中最基础的组成部分。所以如果我们将桥梁看作是一个整个的人体的话，那么静力学状态就仿佛是其心脏。当然，身体还有其他的部分，而在桥梁上你也要考虑其他的方面。整件事情的关键在于如何解决力在两个堤岸之间进行传递的问题。

我最早设计的一座桥，是早在瑞士的苏黎世高工（ETH）的毕业设计。这座桥的设计灵感来自于两个截然不同的想法的组合，其一是悬臂桥，其二是位于其中的一段应力沿拱的方向释放的优美拱桥。在悬臂状态下，桥的受力状态与拱不同。它上部受拉而底部受压，这样一来，一部分扭曲的弯矩就转化为竖向支撑力。我开始从悬臂桥入手，将起支撑作用的部分与其分开，以分别表述这三种力所产生的影响。因此我发现，在桥板下保留一小段空间非常具有新意。

至于竖向的支撑，最初的构思始于我联想到一些与人体比例相关的某些柱子的造型。事实上，从前面看去，这些柱子的确有些像将双手举过头顶的人的身体，它们支撑着桥面。而位于较低一些部位的构件，又使人想到在进行辅助支撑时的人的头顶。或许，这听起来有些矛盾，我在设计时没有加粗竖向支撑的悬臂截面，却反而缩小了构件的截面尺寸，使它看起来与桥面完全不同。这是因为桥的横截面是由道路所需要的宽阔尺度所决定的，而支撑桥面的悬臂截面大小，却完全根据它所受的压力进行计算而得出的。

我非常荣幸地向大家介绍由我自己主持的第一座桥——塞维尔的阿拉米罗大桥。为了举办 1992 年的世界博览会，我打算设计两座对称的桥，一座位于岛屿的一面，另一座位于相对立的一面，再由一条高架桥将它们连接。首先，我从高架桥的构思开始——它应具有许多不阻挡视线的支撑点，以及与当地环境相对应的尺度。这样，桥的许多根吊索也都是针对上面的思考而做出的应对措施。

但是由于其他一些原因，我只完成了其中的一座桥的设计。

它非常具有独创性,据我所知,以前从没有人设计过此类桥梁。通常斜拉悬索桥的侧向拉力必须由塔墩另一面的拉索去平衡。但是当你将塔设计成斜向时,索的拉力就部分由塔的自重所抵消。如果塔的斜角与重量恰到好处,就基本上可以完全承受整座桥梁的荷载,这就是我在此所追求的。

  从机械施工的角度来理解这个项目尤为重要的,我此前在谈及这座桥时,也不只一次地说到这一点。大家仔细想想,每段塔都有向下倾覆的力量,而与张紧的悬索连接后,它们的合力正好与沿桥塔的方向形成对塔的压力。而其他每段的塔都有自身的重力,又同时受拉力作用——如此这般,每段的合力都沿桥塔内部向下传递。最后,作用力与来自桥面的水平力又结合,使力到达基础时转变为完全垂直向下的荷载。基础在与桥身巨大的跨度相比之下显得非常细小。有意思的是,当永久荷载的作用使合力指向一点时,活荷载的作用使合力产生偏离。于是我们遇到了另一个问题,还必须在悬臂系统下考虑风荷载和其他活荷载的作用。

  现在让我给你们讲讲我的故乡巴伦西亚和城内的图里亚河。巴伦西亚至今还遗留下许多珍贵的石桥,在我看来,世界上几乎没有其他城市在这方面可以与之媲美。这里有许多哥特早期、晚期和文艺复兴时期的拱桥,毫无疑问,它们如同这里的建筑一样,都是我学习和借鉴的对象。事实上,如果你亲眼看到旧城门和通往这些城门的哥特式桥梁,你就会发现城门的塔楼与桥头堡在建筑语汇上的差别简直微乎其微。这里还有一座桥梁，其上不仅具备了巨大的公共室外楼梯，而且还承载了一座小礼拜堂；而另一座建于20世纪早期的混凝土桥上，有女象柱和类似的其他装饰物，那是由特瑞尼科——该城的一位雕塑家所做的。这证明了如果时间稍稍往前追溯一点，那时的建筑设计与桥梁设计是紧密结合在一起的。桥梁的建造者对此种特殊关系早就有所认识。他们对建造桥梁的技艺也同样十分在行。这些桥中有许多已长达五六百年的历史，经历过多次难以置信的洪水的冲击。但是即便如此，在施工过程中建造者仍然考虑了桥梁的需要，坚持在桥上建造了小礼拜堂（或许是为了在里面进行祈祷，不要被下一次洪水冲跨）。

　　基于这种文脉，我在旧图里亚河道上设计了一座类似的临时性的拱桥。在它的下边有一个广场，广场下是地铁车站。它如同其他石桥追求材料的纯粹性一样，是纯粹用钢材建造的（我在巴伦西亚城外还建造了另一座桥，除了几根特殊的构件外几乎全部是用混凝土建造的）。在这座桥的设计中，我将其设计成两部分。这首先是因为它必须非常宽以满足交通需要；其次是它与一条位于桥的一侧的林荫大道相接。为了与林荫道的空间相呼应，我在设计时使桥的中间空出，在桥下的空间里，我试图创造出一种空旷的效果。在低洼处建造了一个水池，将阳光反射到桥的底部，水中还映射着桥的倒影。在步行道与公路之间的开口处，阳光从顶部和侧面进入。

下面介绍的是为巴塞罗那市设计的菲利普二世桥。在这里我要特别向大家强调的是，在这座桥中体现了诸如桥梁等公共设施在施工过程中是如何影响城市的地段环境的。这座桥所在的巴赫·德·罗达地区，是当地的贫民区——简陋的棚户区。随着这里居住的人们向环境较好的住宅区的迁移，该区很快与桥和火车站相接，并在它们的下面形成了一系列公园。所以在我看来，巴塞罗那的许多简陋地区，都存在变成这座城市的一处标志性的亮点的潜力。在这个例子中，建造一座桥梁或是满足联系两片地块的交通需求，直接导致了城市的地段更新。

当初设计这座桥的一条重要思想就是建造一个场所。我设计时将桥的中心加宽，仿佛一个戏台一样，横跨铁路的拱不仅标志了桥的存在，而且炫耀着它是这片地区的中心。当初这一地区的景色并没有丝毫的浪漫可言，但从另一个角度看去，尤其是当我望到远处的山峦和周围的民房时，我意识到它非常具有潜力，能够开发成为巴塞罗那的一处胜景，尤其是当现在桥的两边绿树成阴之后。

我认为所有的城市在进行更新改造时，都需要新建公共设施，现在更应如此。20世纪70年代，建筑师与城市专家们专注于从历史的角度出发，将城市的中心区改造得更为适宜居住。这种专注使得许多在房地产商业投机中濒于被毁的老建筑得以保存。但在那时，城市周边地区的发展更为引人注目。于是现在，我们把目光重新投向如何改造这些城市的旧城区——尤其是在欧洲的城市，那里的居住人口在过去的三四十年中增长了一到两倍。我认为诸如桥梁与车站这样的大型公共设施，都会在城市更新改造的过程中因其为城市创造和增添的新的活力，而具有非同凡响的意义。

下一个项目涉及有关拱的应用。第一个实例是位于加纳列群岛的特纳里弗商业展览大厅。有时在设计纯功能性的建筑如许多桥梁时，必须确保造价的低廉，因此如何应用拱就成为一个重要话题。拱在征服大跨度时非常有效。在这栋建筑中，有许多种式样的拱：混凝土三铰拱，还有顶部的钢拱。大跨度的拱可形成240米——约800英尺的净空。

我非常想着重说明一下斜拱，下面我给大家介绍的第一个例子是拉·德维萨步行桥，建于西班牙加泰罗尼亚北部的一座风景优美的城市里波尔。在这片美丽的胜景中，我们希望架起一座桥梁，将特尔河一侧的街区与另一侧的火车站联系在一起。车站与街区间的地段改造成为一个公园，其中包括有一个广场和一系列由其他设计师布置的景点。如前所述，当你想更新一处地段的景观时，桥梁的作用非常明显，它们可以成为更新周围环境最好的理由，并且在建设过程中使失落的街区再一次焕发出生机，拉·德维萨步行桥便是如此。

在此我将着重介绍桥横剖面和斜向的拱的构造，因为当仔细研究了扭转现象之后，我在拱桥的设计上采用了大胆的非对称布局，并取得了一些突破。在里波尔桥的设计中，由于承受荷载的构件拱偏向桥的一侧，桥身的荷载与自重在支撑点上形成了扭矩。这种扭矩传递到一个与桥面同长的管状构件中，由它将桥面与斜拱连接。在巴伦西亚桥中，我们将这个管状结构转化成了整个桥的截面，所以桥面本身就存在抵御扭曲的功能。在安达罗亚海港大桥设计中，我们设计了一个侧面的箱型梁架和一个向外悬挑的步行走道，从步行道伸出许多固定拱身的肋，用它们抵御扭矩的作用。从拱身又伸出许多构件拉着侧面的箱体。在 orléans 桥中，扭矩则完全由道路的截面承担，拱的作用只是通过悬索吊起桥面。

  对拱桥的突破是从美里达桥开始的，这座桥的拱在道路的中心，它防止着桥面扭矩的产生。这是一个由三条纵排腹杆组成的巨拱，它是最早的成功实例。第二个例子是里波尔桥，它是我用斜拱征服 70 米亦即 230 英尺跨度的尝试，它造价低廉，步行道只有 3 米或 10 英尺。第三个例子是安达罗亚海港大桥，它需要承受很重的荷载。第四个例子是巴伦西亚桥，它的桥面由四车道组成，而且两边都没有人行道。

  设计这种类型的桥，最有意思的是对扭矩的处理，在许多标准的拱桥的纵剖面中，我们知道拱是垂直于地面升起的，桥面没有偏心荷载，于是根本没有必要去抵御桥身路面所形成的扭矩，但我在设计这些桥时却充分考虑扭矩的作用，抵御扭矩后形成的不对称的布局，使我能够强调桥梁与周围的城市之间，或者水流的方向，甚至是太阳的方位的关系。它使我更加自如地去把握桥的造型，使之融入周围的城市景观中。

安达罗亚风景如画，面临蔚蓝的大西洋，有美丽的港湾和渔船。在进行波多桥的设计时，我试图利用当地的各种地理条件和建筑材料，比如，用当地的石材来建造堤岸。许多人都从这里漫步到安达罗亚惟一的海滩，所以应该为人们提供一条惬意的道路，因此我决定在桥的一侧建一条宽敞的平台。桥的另一侧还设有一条步行道，但它的主要作用是用于人流与车行道的分离。在吊挂着的步行道旁，我设计了拱的支座的位置，拱上的吊索承担着桥的重量。在巴伦西亚桥的设计中，我同样对桥下的空间产生了兴趣。安达罗亚的潮水变化多端，许多人驾着小船在桥的周围，因此我将堤坝建造成直接通向水边的非常实用的大台阶式样。我将桥体分开以使光线进入桥的底部，它与堤坝一同形成了非常有趣的空间。

建于毕尔巴鄂的坎普·瓦伦蒂尼桥，其特点在于它与我们所学到的或想到的传统桥梁都有所不同，尤其是反映在堤岸的形式上。当你想到传统的桥梁，比如威尼斯的桥时，会发现它们的拱都将所有的荷载直接传递到岸边坚实的土地上。而这里却采用了阶梯状的桥堤，人们可以拾级而下直达运河。桥堤是桥梁永恒的组成部分，它反映着桥梁是如何通过与地面的接触而将力延伸出去的。

在毕尔巴鄂，我设计的桥由沿着河岸两边与河岸平行的悬臂体系所支撑着。桥身就架在这些像手臂一样的支撑体上面。悬臂体系的剖面呈半拱形，取消了实体的桥堤，从而形成了中空的部分。支撑体与中空部分都形成了桥的方向感，人们可以看出它们与水流的方向一致。

步行的流线先是与河流平行，接着又交叉跨过河面，而后又折回到最初的方向。我用一种非常简单的方式暗示这种路线——在桥的平面上设计了一道曲线。于是传统桥梁的严整直线特征消失了。我想设计一个构件，用来承受这种同样呈不对称状布局的结构，以在视觉上与桥身的不对称性相呼应。于是拱偏向一方与曲线配合。

桥面下用于抵抗扭转的管状物是直线延伸的，所以设计管状物与拱交点的位置要十分精心。因为拱的受力在剖面上呈对角分布。管子中的扭曲弯矩在水平方向传递着，受力的状况使得还有几处需要竖向支撑，它们设在桥梁几处反弯点的位置上。拱的支撑点设在距离它与抗扭的管状物三四英尺处，虽然从荷载的角度看它还是受轴向力。我试图清晰地表现我对该桥受力状况的清晰认识。比如，从平面看去，你会发觉扭矩是通过类似球状的体系抵消的。抗弯的管状物一侧的桥面与另一侧的面积相等。最后，我们知道管子两边的扭矩相互平衡。

今天我们或多或少失去了某些20世纪60年代那种用建筑解

决社会问题的信念。实际上，我们今天的确忽视了一些社会问题，尽管还有三分之一的人每天不能够很彻底地解决温饱问题。我的意思是，这十分富于戏剧性。你可以想像仍然还有多少基础工作要做？想像如果追溯到崇尚机器的时代，那时人们到达一个新的地方后，他们去将水引到栖息地、去拦截湍急的河流，或者修建卫生设施时有多么困难。我想当你设计桥梁时，也会同样强烈地感受到这些困难的存在，尤其当你的桥是伫立在城市之中的。

我仍认为桥梁与桥梁设计的潜力尚未完全发挥。桥梁的活力不仅仅在于它的交通作用，更在于它是城市中无可非议的具有标志性的重要组成。你能想像出纽约失去那些美丽的桥会是什么样子吗？或者华盛顿桥变成多跨桥，而不是像现在这样一跨横跨足足1000多米、超过半英里的空间的形象吗？还有将它的主跨变成现在的一半或更小又会怎样呢？然而通过工程设计，却可以驾御这些情景中那些充满野性与强劲的因素。

如果看一下19世纪的桥梁，你会发现工程师对扶手和灯具的设计也同样留意。他们知道对细部的推敲可以赋予桥梁更多的益处。有些人说道："建筑是将之拿掉之后仍然使桥梁屹立的东西。"此言诧异。建筑就是桥梁本身，因为桥是为人而设计的，所有满足人们需要的设计都对桥梁的设计有益。

运动与形式

今晚,我将对大家讲述第三个议题,或许也是最困难的话题,因为从某种程度上来说这对于我是最为保密的工作。它是建筑的构思、模型和图纸累积而形成的结晶。跟大家讲述这些,就如同厨师向世人——尤其是青年人泄露自己的秘方,即用什么原料、什么调味品来形成这样或那样的味道。虽然它看起来天然并具有个人的色彩,但却是我在学术生涯中领悟的最高境界,体现了如何将不同的艺术手法、艺术流派、技术因素以及建筑风格融合在一起。本想长篇大论地讲很多的议题,由于这些,我只能简单说上几句,因为它们是直接沟通的最简单的方式。

在前两讲中,我都是由孩子玩具构成的小玩意开始的。它们都很小——统统不足一英尺宽。两个雕塑都涉及如何悬挂物品的机械。如果你仔细观察,就会发现其中惟一的不同就是荷载的布局有所差异(我特别用该词以示工程用语),使得受力结构发生了变化。在第一个实例中,构件的受力是平行的;而第二个实例则是垂直的,但它们都只有一个目的——悬挂石头。关于这些,我要强调几点。首先,悬挂石头本身是非常重要的,因为你在与使它下落的趋势作斗争。就像牛顿被苹果砸中了脑袋后所发现的,我们生活在一片充满力与引力的土地上。如果作用力与引力不是如此作用,那么也许我们的身体都会完全不同。

其次,小构成之所以紧紧连接在一起,也是由于石块的重力作用。这意味着,如果那里没有石块,这个构件将很难固定在一起。于是这种重量或是持续的引力也是大自然的馈赠。这个构成之所以保存,是由于引力的方向没有改变。如果我们切断传递重量的连线之后又会怎样呢?非常明显,石块会落下,而紧接着整个系统将崩溃。当石块落下时形成了一种运动,整个系统的静力结构却都与形成这种运动的趋势相关。这意味着受力的结构总是与物体的运动相关,没有运动就不能形成受力结构。在我看来,这是一个完美的解释,因为即使在静态的受力结构和非常稳定的体系中,也隐含着运动的趋势。正是通过运动的存在,使我们产生了力与形式的概念,我在前面的讲座中曾经阐述了力与形式之间的关系,但是运动与形态的关系却实在难以用幻灯片来表达。

我希望从我在苏黎世高工读书时所做的两项工程谈起。第一个项目是在于尔格·奥尔瑟的指导下与一群学生共同完成的。我们在苏黎世高工的塔上建造了一个游泳池。它是由 24 个钢索,悬吊着 24 个肋上的 1.2 毫米厚的聚酯构成的。可容纳 24 立方米的水,甚至可以在里面游泳,因为脚可能会踩坏装水的膜结构,所以人们不能在水中呆得很久。这个结构还有一个挑战,那就是它建在塔底,游泳池的正下方就是图书馆,你能想像一座图书馆的上面有那么多水吗?

与这个项目有所不同,但却并非全无联系的是下一个项目。前一个项目注重的是体现悬吊很重的物体,这一次却为的是研究多面体的清晰运动。我的博士论文是"结构的可折叠性能",它就基于某种多面体的拓扑关系研究，那是一种复杂的结构，可以被折叠成一束，使所有的杆件都相互平行。经历一系列的变形过程后，束状体展开，慢慢复原成多面体，形成与穹顶类似的形体。尽管它看上去并不完美，但它揭示了如何从束状展开成为半球状的复杂过程。

我要强调的另一个产生趣味的源泉，是自然和对自然直接的观察，它意味着去直接观察我们周围的自然的存在方式。你要去观察花草树木和一切自然中的事物。从许多对骨骼的研究中，我早先制作了一个模仿树木结构的模型，并想用钢和玻璃来形成一片树林的概念。我不再按照物体本身的特性向下发展，而是通过一套体系来模仿原先的物体，我决定使它们可以变形，在这里意味着它们自己产生变化。于是树的顶端的枝桠形成了屋顶——铰接于支撑它们的部位，整个屋顶在机械作用下变形张开。这种树的概念后来在多伦多的一个画廊中得以实现。但机械变形的想法却没有与树结合，而只是形成了一片很大的、可控制开启的窗体，它与静静地伫立在遗迹广场上的画廊分开，形成了喧闹活跃的空间。

我们设计了里斯本的东方火车站，在我看来非常自然地沿袭了树的想法。在这里我又一次运用相同的建筑语汇，这不仅是由于里斯本城市本身的美丽，还有与它靠近的柔媚多姿的大西洋海港风貌。该空间在美丽的阳光下显得异常清澈。基于这种文脉的考虑，我不想采用像里昂火车站那种非常粗壮的结构，而偏爱更为轻柔的、开放的结构体系，人们可以在建筑的一面看到建筑另一面火车的穿过。

建筑另一个同样非常重要的主题是解剖。设想人们在如同人体般的建筑物里活动，或是在如同人体一样的建筑中进行鉴赏。不管怎样，事物的尺度总是与我们的身体有一定的关系。建筑也非常自然地与人的尺度相关，因为它本身就是为人而建造的。这使得解剖成为创作灵感的一个源泉。不仅仅是文艺复兴时期，人们将解剖学作为确定尺度与比例的基础，直至20世纪也以它为模数。解剖学中，张开的手掌的形象、眼睛的形象、嘴和骨骼的形象都是灵感的源泉。通过研究我们身体的结构，你可以发现一种对建筑非常有益的内在逻辑性。

认识建筑中的造型与雕塑特征是非常重要的，它并不会与建筑的功能性相悖，也不会与结构发生矛盾。比如在里昂的车站的翼状屋顶，就是研究了眼睛的结构后发现了其中的几何雕塑规律的实例。

同研究眼睛相比，我更加抽象地去研究了人体的头部如何固定在身体的上方，以及如何或怎样进行扭转？因为整个头部都支撑于脊柱的第一节颈关节上，所以它可以自由扭转。这使得抬头、转头、低头或组合起来的动作，在某种程度上非常有趣和富于戏剧性。我曾经研究过如何控制头部的运动，开始是从纯体量入手

——用最少的构件固定立方体。比如用非常细的细杆和围绕在它周围的一些线。在另一个雕塑中，那个被看成头颅的物体在垂直方向上只有一个构件支撑，另一个斜向的构件起到拉接作用，将其固定在水平位置上。在斯塔德荷芬车站的剖面设计中，运用了同样的原则，但是方块状的形体被车站的体量所替代而整个雕塑的位置就伫立在大地上。

我非常欣赏简洁纯粹的构思，就像简明扼要的符号表达有力的事物一样，复杂源于清晰的思路中的多种思想。这意味着虽然每个思想都可以自成体系，你也可以从中将几种思想分出主次，如同不仅使用黑色还辅以其他几种色彩素描，或者为了控制水平用蓝色绘制辅助线的画家一样。所以在斯塔德荷芬车站设计中，除了模仿头部的支撑外，还借鉴了张开的手的概念。这种模仿贯穿于整个项目里主要的扶壁柱、一个小顶棚、候车的棚架之中，使它成为车站大部分建筑外观的一大特征。剖面底部地下空间的柱子，是将手的形状倒置的形式。这种几何造型还可以在建筑的许多部位继续引用。

另一系列对雕塑性的决定性研究取得了更深的进展。这里描述的是脊椎构造的研究，揭示我们的身体如何得以直立的秘密。脊柱是由许多脊骨构成的，我们可以在雕塑中用非常自然的方法表示它们——一连串的立方体。

一旦你有了脊柱的概念，将其中的元素在周围移动就非常方便，可以通过多种方式来改变脊柱的形状。有一次我将立方体绕着中心的绳索周围错落摆放，还有一次它们逐渐出挑以使运动的概念更加明晰。尽管使用的都是同一种材料，但却不显单调，于是你就会对弯曲的脊柱有更多的想法。同样值得一提的是脊柱如何完成扭曲、绕着轴心旋转、弯腰和延伸。有些从前非常含混的运动方式逐渐变得清晰了。

在后期的一些研究中，我将方形的几何体进行变化，以试图获得更高大的雕塑感。在一个项目中，它成为了12米也就是40英尺高的体量。我试图创造一些更为有机的形体，于是我摆脱方形的体量，用两个金字塔的棱柱和八面体来替代它。这里有一组7个的混凝土制成的八面体，它们一个接一个地排列着，每个体块都靠2颗钉子和1根绳索与下一个相连。7个体块通过两个非常长的支座维持平衡。在这种尺度下，这样的雕塑已经超越抽象的几何构成，而完全遵循了结构的原则，它成为我设计塞维尔的桥的基础。

从单纯和抽象地研究人体和重力解剖关系入手，人们可以推想钢材的主要受力情况。当你开始建造一个12米高的雕塑时，将它搭建的过程就成为一个实实在在的事物。怎样固定物体、如何选择材料，这些问题都非常关键。你必须采用混凝土或者钢索。你必须考虑支撑和拉结，还有运输与其他的方方面面的问题。

从一个特殊的视角审视塞维利亚大桥，你会发现该桥具有非常抽象和原始的桥梁的概念。但当从前方或后方观看桥梁的时候，你不会想到它是桥。相反，它表达了某种自由的意境。一个塑造或雕刻作品的过程本身就非常美妙，因为你感觉到了自由。你只是被你自己所设置的标准所束缚，比如，"我想要传统一些的，我只考虑用方形体量"等。不管在怎样的特殊状态下，你都会因自己的表达方式而受到局限，但你仍然非常自由。因为你的惟一目标是追求一种纯粹的雕塑感。当你处理桥梁或者建筑，甚至是简单的雕塑时，都要考虑功能的需要。而此外，你还有很大的优势可以利用，那就是尺度，没有任何的雕塑能够与桥梁和建筑的尺度相比，这就使得建筑尤其是与工程相结合的建筑具有非常的意义。

　　在这里我非常想与大家讨论一些与运动相关的内容，但主要是针对于可以变化的结构单体。在"塞维利亚世界博览会"上，我设计了科威特的展亭，运用了一系列的、可以开合的半拱结构。这个构筑物覆盖在一片由半透明的大理石铺砌的露台上部。露台下方是亭子的内部空间。白天阳光透过大理石照亮了下面的室内。组成屋顶的构件由混凝土支撑着，每一个都装上了马达，以使屋顶可以慢慢地张开和变形。由于每个构件都是独立的，所以你可以自由控制屋顶的开启方式。在屋顶开启的过程中，人们会联想到手掌和手指弯曲的动作。它们仿佛在保护着里面的空间，当手指张开时，又像在拥抱天空。

在纽约现代艺术博物馆的庭院里垂柳旁边,我们设计了一组雕塑。垂柳的枝条弯弯而雕塑也缓缓向下弯曲,每个都由一个轮子控制,几乎要触摸到下面水池表面的由 aristide maillol 设计的名为"河流"的雕塑。

从一系列根据运动的规律所设计的雕塑的观察——特别是由围绕不同圆心旋转的直线生成的拓扑曲面状的形体,引出了另一些对屋顶结构的探讨。在这些雕塑中,直线生成了折叠的弯曲表面。在接下来的研究中,这些直线又成为单独的建构构件。先是在地面上得到一个形,然后是位于中心的斜向的脊线,接着再在上悬挂着许多同样的构件,组成形体的一个剖面呈圆形,而另一个剖面呈半椭圆形。这个研究同样成为探讨如何使这样的形式运动与开启的基础。两部分结构在中心的连接处,仿佛是两只紧握的手在拇指处交汇一样。它们环绕铰链的轴作开合的运动。

在我们现在设计的密尔沃基艺术博物馆扩建工程中,将这种屋顶结构进行了变化。在我的方案中,我将埃罗·沙里宁设计的老馆和大卫·卡莱尔设计的新馆连在了一起,建造了一些似乎是属于 20 年前城市中的桥的形式。博物馆本身就有些类似桥的形式,我在设计中十分尊重博物馆原先的构思和它与城市的关系。这座博物馆现存的建筑中,靠近湖面处已经建有一座桥和一个雕塑般的造型,而我设计的方案中又重新使湖的前方拥有另一座桥和雕塑造型。这些形体靠得很近,于是新设计就构思成穿透式的方案。

以一串轻盈的棚架连接了新老建筑，而从桥的高度上，人们可以一眼望到湖水的尽头。

在建筑设计方面，我还想向大家介绍从来没有完成的圣约翰天主教堂。它最初由海因斯和拉法热设计，继而由罗夫·阿达姆斯·克莱默接手，还有一些拱顶的部分是拉菲尔·古斯塔维尼奥设计的。现今只留下广场与教堂半圆形的后殿。在教堂入口的竞标方案中，我尽力从以前大家见到的建筑语汇中创造出一些象征性。因为象征性的语汇在天主教堂建筑中都非常显眼。我将天主教堂想像成一株大树，底下是它的根和树干，然后是上面的枝叶。在竞标任务中有一项任务是创造一种"有生命的遮蔽体"。这种有生命的遮蔽物理应在室内出现，但我希望最好将它放在屋子的上面。在教堂下面与屋顶间的拱形空间，一般都很晦暗和封闭，所以我想将它打开。

我的设想是替换现在已经破损的屋顶，设计一个玻璃的屋顶并在上面植树，在教堂的顶部形成一座园林。园林是整座宗教建筑的翻版。非常有趣的是，在贝多芬的作品中，比如赞美诗、合唱和钢琴曲中，都暗示过大自然的神殿。大自然像是一座神殿，所以引发我们去试图建造自然的殿堂，这是个多么浪漫的想法呀！

天主教堂的平面是依照一个拉丁十字布置的，人们可以将人体比作坛庙，那么在拉丁十字中我们又可以看到人体的概念。所以这花园和人体都隐喻着拉丁十字的几何形状。这种起源很早的神秘主义，也同样是建筑构思的组成部分之一。

我想使园林顶端的玻璃屋顶的开启可以得到控制，从而使园林可以接受雨水的滋润。三角形的玻璃顶的剖面，起到和多伦多展廊的大玻璃窗同样的功能。它们围绕着一个轴进行旋转，屋顶被设计成高高的尖顶状，可以通过它来调节温度，在室内营造一个舒适的小气候。

为了尊崇自然，同时更为了表示对城市景观的重视，我们必须在建筑与工程方面进行思考。我非常愿意对此进行强调，虽然此处无法详细论述，但重要的是将建筑理解为整个城市景观的一部分。我认为在圣约翰天主教堂中，我所能总结出的基本思想是达到了一种非常有趣的平衡状态。

结　语

　　我所设计的这些方案之所以可行，不仅是由于人们采纳和实施了它们，对此我非常荣幸，还由于它们是集体努力的结晶。在我的工作室里，人们制造模型、绘制图纸，大家都参与到设计当中并对方案的发展贡献力量。我妻子对我的这次演讲给予非常大的支持。谢谢大家！

　　这表示演讲即将结束，感谢大家前来参加。

Santiago Calatrava

Conversations with Students

THE MIT LECTURES

Santiago Calatrava: Conversations with Students

THE MIT LECTURES

Santiago Calatrava: Conversations with Students

THE MIT LECTURES

Cecilia Lewis Kausel and
Ann Pendleton-Jullian, editors

Department of Civil and Environmental Engineering
Department of Architecture
MASSACHUSETTS INSTITUTE OF TECHNOLOGY

PRINCETON ARCHITECTURAL PRESS

Published by
Princeton Architectural Press
37 East Seventh Street
New York, New York 10003

For a free catalog of books, call 1.800.722.6657.
Visit our web site at www.papress.com.

Material presented in this book is also available at
http://web.mit.edu/civenv/Calatrava/.

©2002 Massachusetts Institute of Technology
All rights reserved
Printed and bound in the United States
05 04 03 02 5 4 3 2 1 First edition

No part of this book may be used or reproduced in any manner without
written permission from the publisher, except in the context of reviews.

Every reasonable attempt has been made to identify owners of copyright.
Errors or omissions will be corrected in subsequent editions.

Editor: Nancy Eklund Later
Designer: Deb Wood

Special thanks to: Nettie Aljian, Ann Alter, Nicola Bednarek, Janet Behning,
Megan Carey, Penny Chu, Jan Cigliano, Russell Fernandez, Jan Haux,
Clare Jacobson, Mark Lamster, Linda Lee, Jane Sheinman, Katharine Smalley,
Scott Tennent, and Jennifer Thompson of Princeton Architectural Press
—Kevin C. Lippert, publisher

Library of Congress Cataloging-in-Publication Data

Calatrava, Santiago, 1951–
  Santiago Calatrava, conversations with students : the M.I.T. lectures
/ Cecilia Lewis Kausel and Ann Pendleton-Jullian, editors.
     p. cm.
  ISBN 1-56898-325-5 (alk. paper)
  1. Calatrava, Santiago, 1951—Themes, motives. 2. Architectural
design. 3. Architecture—Technological innovations. I. Title: Santiago
Calatrava. II. Lewis Kausel, Cecilia. III. Pendleton-Jullian, Ann M. IV.
Title.
  NA1313.C35 A35 2002
  720'.92—dc21
                                 2002003635

CONTENTS

PREFACE
RAFAEL L.BRAS AND STANFORD ANDERSON   52

INTRODUCTION   56

MATERIALS AND CONSTRUCTION PROCESS   58

FORCE AND FORM   94

MOVEMENT AND FORM   126

CONCLUSION   156

PREFACE **IN NOVEMBER OF 1995,** I was taking a sleepy walk in Valencia, Spain, when I saw this unusual but simple and very beautiful bridge. I inquired as to its architect, and the immediate response was "Santiago Calatrava, of course." I must admit that, at that point, I did not know who Santiago Calatrava was, but I am a quick learner and immediately proceeded to remedy my ignorance. At my request, a colleague, Professor Herbert Einstein, contacted Dr. Calatrava and began discussing ways to get him involved with MIT. It turns out that Santiago had already been invited to visit the school by the Department of Architecture. That visit gave us the opportunity to meet and to organize a series of lectures. This book, and its accompanying Internet site (http://web.mit.edu/civenv/Calatrava/), document the extraordinary exchange that occurred between Calatrava and a large audience of students and professionals over three days in 1997.

Hearing Santiago Calatrava speak reminded me of why I wanted to be a civil engineer. It reminded me of my own desire to create, to design solutions that are functional and beautiful, to leave behind works that will be remembered. I suspect that all civil engineers—and all children—have the same dream; unfortunately, our educational system conspires to dampen it. The idea of the architect-engineer has been lost. Creativity is buried under equations or hemmed in by the walls of specialties.

Calatrava represents what the architect-engineer should be. His bridges and public buildings reflect a

THE DIVORCE OF ARCHITECTURE and engineering is long standing and now, at least in the United States, almost ubiquitous. This divorce injures both parties. The ambition of architects to build well is diminished. Engineering becomes formulaic and uncomprehending of its social, environmental, and aesthetic dimensions.

As a school, MIT cannot exist independently from the conditions of our time and place, but we can foster ambitions to restore a profound alliance between architecture and engineering. Happily, there are those creative individuals who hold such ambitions and create exemplary works. In architecture, one thinks of Renzo Piano and his "building workshop," where building well is manifested even with the special challenges of building innovatively. But for all the excellence of Piano and his shop, he, like many other fine architects, works through a process of collaboration with the all-too-rare creative engineering firms. One thinks of Ove Arup in London, Buro Happold in Bath, and RFR in Paris, all of whom collaborate with architects to achieve works beyond the scope of either partner alone.

Especially in bridges, infrastructure, and long-span buildings, one finds engineers who control the entire design and succeed both in technical and aesthetic terms. Through our collaboration in the Felix Candela Lectures,* we have brought such practitioners to MIT, including Heinz Isler, Minoru Kawaguchi, Christian Menn, and Joerg Schlaich. Each of these engineers would be wholly convinced deep understanding of engineering. Like a classical arch, his structures seem to flow with the forces and, vice-versa, the force vectors seem to merge with the structures. There are no superfluous elements. The motion, real and apparent, of his creations is smooth and effortless, like motions in nature. Indeed, Calatrava commonly finds inspiration in the human body, the most beautiful and functional of all natural objects. His buildings, like the body, integrate individual elements through simple interactions to create enormously complicated machines.

Not everybody can be Santiago Calatrava. Not everybody is blessed with the same talent and artistic sensibility. Nevertheless, every civil engineer can strive to be more creative, and every architect, to be more imaginative and aware of the interplay between structure and mechanics. If we did, our professions would be far more exciting.

Rafael L. Bras
Bacardi and Stockholm Water Foundations Professor
Former Head, Department of Civil and Environmental Engineering
Massachusetts Institute of Technology

of the sound scientific principles their designs embody. Nonetheless we also observe a personal signature in their works.

Santiago Calatrava, architect and engineer, unabashedly pursues the unity of art and science. His exploration of natural forms (particularly of the human body), his readiness to work metaphorically, and his brilliance in representation all facilitate his creative exploration of form, space, light, and even kinetics. His mastery of engineering principles not only allows the realization of his designs but is challenged and advanced by the dialogue between formal invention and scientific principles.

The genius of this process is embodied in the energetic manner of Santiago Calatrava and powerfully advanced by the drawings that inform his lectures. We hope something of this dynamic survives in the static form of this book and will contribute to the larger ambition of promoting the fruitful common bond between architecture and engineering.

Stanford Anderson
Professor of History and Architecture
Head, Department of Architecture
Massachusetts Institute of Technology

*A series of annual Felix Candela Lectures was launched in 1996 by the Structural Engineers Association of New York, the Museum of Modern Art, and the departments of architecture of Princeton University and the Massachusetts Institute of Technology. Beyond honoring the creative achievements of Candela at the frontier of architecture and engineering, the series recognizes such excellence in current practitioners and seeks to advance these concerns through education.

INTRODUCTION

Ladies and gentlemen, I thank you very much for the opportunity to speak here in this school. After having been a student for a long time, studying in Valencia and then in Zurich for something like fourteen years, I started my practice as an architect and engineer. For sixteen years now, I have been working very intensely in this practice, and the only contact I have had with institutions like MIT is in sporadically giving talks. This is the first time that I have made a commitment to give a series of talks with the specific intention of communicating my experience. I think it makes sense now because these sixteen years form an important period in my life and because they define a generation—one stage in the life of a person. The things that I am saying I say for the next generation—the people who will look at my work and invent other styles and find their own way, just as I have integrated the work of those before me in finding mine.

I thought it best to speak about my own experience because, in fact, this is the only thing that I know. I mean to look back at the works that I have done and try to introduce you to the very essential thoughts that have informed my work during those years and the steps that have permitted me to go from one building to another, trying each time to implement a little bit more of my thinking.

MATERIALS AND CONSTRUCTION PROCESS

I choose to speak first about the idea of materials, because it seems to me that, in terms of architecture, materials are fundamental. After all, in architectural ruins you find only stones. So the material part of architecture—let's say the physical support of the architecture—is, in my opinion, very important and very fundamental.

I thought that a good first step toward understanding architecture was to understand what concrete is, what steel is, what wood is, how to use them, what they signify. What are the forms they can achieve? What are the differences between these forms? This is what I want to try to show you, beginning with the very first projects I undertook and continuing with the most recent works.

In this small figure done by using pieces of toys, the force of the hanging stone goes through all the pieces and is materialized in stone, wood, cord, and parts of steel. It seems very simple, but there is a lot of complexity there: the different pieces of the toy working in tension; the spindle that separates the parts working in tension, itself working in compression; even the colors that are put together in a simple but deliberate manner. There is no doubt that the simple act of holding a stone in the air can be a matter of expression.

The work that I did for the Wohlen High School in Aargau, Switzerland, required me to make a series of interventions in some existing buildings. I added an entrance, a central hall, a roof for the library, and another roof for the great hall. In this project, I changed materials several times. One part was built in concrete and steel; another part, in steel and glass; another one, in wood and concrete. Along with the experimentation with materials—using particular materials for particular solutions—I also introduced another theme. I thought that it would be interesting to work with the light, controlling it differently for each particular space.

The idea of the entrance was generated from the existing plan and its geometry. The plan was a trapezoid, which I cut with a diagonal to create a canopy that consists of two cones attached by an arch. One works in one direction and the other, in the other direction, with a pipe in cross section that provides torsional resistance and also holds the gutter. Even though the pipe has torsional stiffness, I used it here for the purpose of creating a link between the façade and the canopy, so that these elements work together in the same gesture. An ensemble has to become a single thing.

And of course, independent of the fact of construction, it is clear, looking especially at the elevation, that there is the idea of a leaf, or a palm. A very figurative idea was part of the design. There is certainly behind my very first approach—this free approach—to architecture a looking for inspiration in natural forms. The simple observation of things motivates me as much as the material aspects of architecture do.

The second intervention at Wohlen High School is an entry space. There I made a circular cupola in wood. The shape was very straightforward; a circle was subdivided into radial segments. These segments were made using a v that I cut, opening the interior. I replaced the portion of the crease that was removed with a linear element that signifies the compression in this element. I separated the different components of the cupola's support in order to make visible all the different types of support and to create a free ring around the periphery. This tension ring is floating. The idea of defying gravity is expressed in showing this ring—making it visible but not structurally significant—and then pushing back the corner supports of the v segments, which are the real supports of the cupola.

The light comes in from behind the cupola, at its base and through the creases of its triangular segments. The surfaces behind the cupola disappear in this light, so that you get the impression that the whole thing is floating in the interior space. It is clear, when you see all those shapes together, that they can be associated with petals.

In a library, the control of the light and the relationship of the light to the space is even more important. I thought it necessary to liberate the walls and to create tangential light, as Louis Kahn might do. I decided to make the roof at the center of the space seem to float. Its main support is a column toward which the roof inclines and through which rain water is channeled. The roof is a shell composed of several shells. The corners are held in place so that the roof will not move laterally, but all the weight is supported through the center column. Then light descends along the walls, tangentially.

For the genesis of the idea I started thinking about a book—an open book. Again, as in several other cases, the idea of the roof as floating came to mind. This is a theme in which the idea of lightness is embedded, and lightness is often created because of the contraposition of materials or of static systems. If the roof is heavy and the spindle is light, this contraposition of two materials combined with the light coming from the sides will make the whole roof seem to fly. The basis for the form of the shell was not the usual hyperboloid or paraboloid; it was a book—an open book. At the same time, the shell also wanted to become a bird. It is a kind of superposition of concepts. You can also see a leaf in it, held in the spindle support.

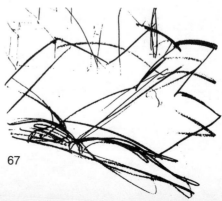

With the design of the great hall, or auditorium, what I wanted to achieve was quite simple. I proposed making the roof emerge by creating a parabolic arch on the interior that supports a raised shell that is independent of the walls. At each side of the shell is a gutter and a longitudinal window that brings light to the interior, very softly emphasizing the underside of the shell and bringing transparency to the repetitive, oblique elements that are transferring the weight of the roof to the arch. These oblique elements are all standardized pieces of wood approximately three inches by three inches. The parabolic arch and an upper arch supporting the shell are of laminated wood. The space is very intimate, partially because of the decision to use wood. The contrast between light and pattern on the interior also contributes to this intimacy.

As in the entrance canopy, there is the idea of a palm tree. Also—and this may be very figurative—I was trying to express the idea of the force of the parabolic arch transferring to the column. Many people think the column top is a reminiscence of an Ionic capital, which is not the case. It is more like the head of a ram.

The columns were pre-cast in concrete. I like very much brute concrete—concrete done on site—but pre-casting can be very interesting because of the complexity and freedom of forms that you can achieve. In Valencia, in my mother language, we used to say *formigó*. *Formigó* comes from *forma*. In Spanish, concrete is *hormigón* with "h" replacing the "f" of *formigó*. *Formigó* means material to which you can give form. This is a good definition of concrete. With pre-cast concrete you are very free to choose the shape, the texture, and many other characteristics of the material.

We cast the columns for the great hall in the most economical way. We cut the column in two and cast each half horizontally and then glued them together. This has another advantage in that all the exposed surfaces in the end are finished surfaces; you do not see any of the places in which you have been casting the concrete.

Ernsting's Warehouse in Coesfeld, Germany, is also an early work. It was an exercise in how to clad an existing warehouse and give it new signification. The first decision I needed to make was about the materials, which had to be very economical. So we decided to use concrete and crude aluminum—crude in the sense that it is the standard aluminum that you can get very easily. You can purchase it corrugated or flat—you can get it in many ways—and it is easy to work with.

We investigated how to transform this building, not only materially, but also thematically. The limitation of the material created a significant restriction; the idea of working with a theme—here, the "pintoresque"—allowed us a certain freedom. By "pintoresque" I mean like a painting; each façade should appear like a different painting executed in the same two materials. The material creates a unity, and the treatment of the material makes each façade different from the others.

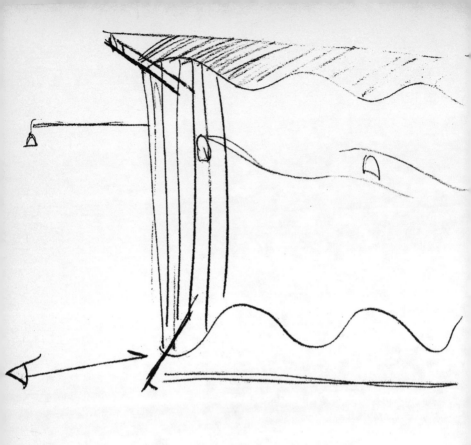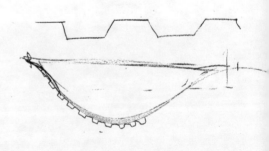

The south façade was done using corrugated aluminum, made into a sinusoidal curve. This gives the façade a double rigidity, because the convex segments of the curve are held at the inside with transversal connections. This curve gives one level of rigidity and the corrugation of the metal itself provides a second rigidity, permitting us to make quite a tall façade.

In order to emphasize the movement of the curve, we cut the façade at the top and bottom on an oblique angle, so that looking from the front, in elevation, the section of the sinusoid is even more accentuated. Looking up to the top of the façade, the curve is clear and readable. The client said that for safety reasons he needed a lot of light on the exterior, so we put lamps on the façade. The light makes the plasticity even more readable. We mounted the lamps away from the façade so that the shadows of the lamps follow the curve. They create drawings on the façade, and this is quite a plastic relationship.

This is the south façade, which means that the movement of the sun during the day makes a very significant change in the façade itself. In the curve you see shadows of the light and also the reflection of the sun. Vertical vibrations are produced by the reflection of the sun in the corrugations of the aluminum. The façade is extremely sensitive, changing with the hours of the day—with horizontal light, with vertical light. The oblique cut at the base makes the façade seem to float and move over the concrete structure.

On the north façade, the problem was different, because it has just zenith light—a very diffused light, ambient light. So how do you emphasize the plasticity of the façade in zenith light? I proposed using an s profile placed horizontally on the façade in very long pieces—as long as possible, which means maybe ten meters, or thirty feet, long. In this profile, the center section, on the diagonal, will generate quite a reflection. At the top edge where the two profiles overlap you will get shadow. Then again, less reflection at the bottom of the profile, until it breaks in shadow again. This makes a linear structure but one that is sensitized to the zenith light.

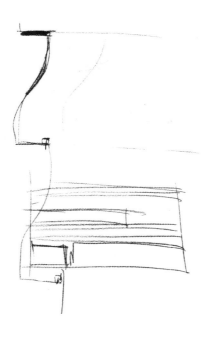

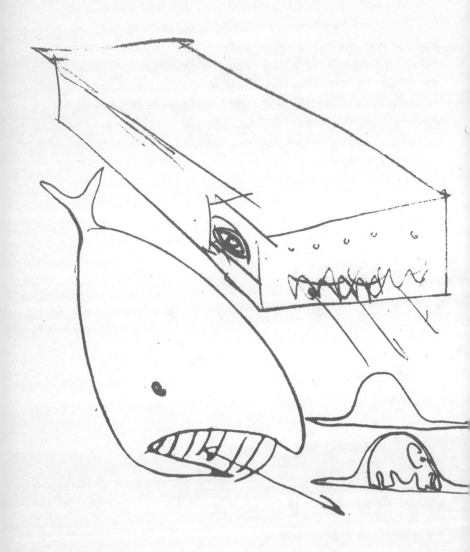

On the east facade we used corrugated plates seventeen meters high. The problem here was to put them on in one piece and to hold them in place with small elements at the bottom and the top. The east side is very flat, confronting the horizontal light of the morning sun. There was an elevator body on this side that needed to be clad. In order to distinguish it from the rest of the façade we used scales—like those of a fish—but very big ones, made of aluminum, that fold at the corners and into the top.

So again, each one of these façades is conceived as a separate picture. I spoke before of the "pintoresque," meaning like a painting. But how do you link these different paintings, these façades? From the very beginning, the building was like a foreign body. We had to find a way to give life to this body. The west façade has almost the same light conditions as the east façade, the sun setting versus the sun rising. It was also done using those very large panels of corrugated aluminum. It incorporates three large gates for the lorries that come in and go out every day. There are a lot of them. Many lorries wait at the door. The gates open and the lorries go in or out. If you look at these doors, at the building's scales, and at other things, the building somewhat resembles a whale. And with the lorries going in and out, it is like the story of Jonas, or like the elephant and the snake of St. Exupery. There is here a theme of swallowing. You see, it is also very important, this kind of game.

The sculpture I showed at the beginning of the lecture was made with toys. It was an exercise for me to take my children's toys and the things in my house—pencils, the cord that hangs the curtains, whatever I found—and make the sculpture. The elementality of the exercise, or of the thought, takes absolutely nothing away from the complexity of the solution. In the beginning something can be extremely spontaneous and simple. I want to walk. Where am I going? That is the question. But, you know, to walk is just a very natural thing, a very simple thing. A very long trip needs to be started with a small step. These thoughts of the whale are, in my opinion, an effort to pass from the "pintoresque" to the textual and to give the whole thing a life.

In the warehouse doors, the module is a continuation of the module of the façade. The pieces of the façade are all very repetitive; only the length of each has changed. In this there is a very beautiful idea. It is the idea that a façade can be transformed. The idea of transformation, of metamorphosis, is a mother of evolution in architecture—an unbelievable mother! It is not only because we can deploy mechanical and physical elements to create new architectures based on the idea of metamorphosis but also because we are maturing in our needs and in our understanding of culture that now is the moment to introduce these components into architecture in a major way.

Can you imagine, for example, a whole façade that gets transformed? Not just half a façade, like in this project, but the whole façade, like a curtain opening from one side to the other. This is, in my opinion, feasible today and certainly a challenge for the new generation and in the evolution of architecture.

The Stadelhofen Railroad Station in Zurich is sited on a hillside. In section, you have the lake of Zurich that used to extend to the base of this hill—a moraine made by glacial deposits. Early settlers walled off part of the lake to create a flat area upon which they started building. Later, the railroad came and made a cut into the hill for two rail lines. The cut separated two areas of very different character—a situation that is still preserved today; on one side it is very urban, and on the other side, very green.

Our exercise was to enlarge the station. It is a station that is heavily trafficked and one that is important within the regional railway scheme. Our intervention considered the fact that there was an existing cut from which we had to step back, cutting deeper into the hill. The basic idea was to conserve the original edge of the cut, so as to be able to retain the green condition above it. For this, we proposed a wall-like structure that is permanently anchored to the hill. The wall supports the houses built on it, which are often very close to the edge of the site. Gardens and a pergola were created along the wall, preserving the character of the upper part of the site and allowing for the possibility of people promenading above while passengers wait for the train below. Beneath the train lines we built an underground arcade for shops, which makes the link to the platforms more safe and more interesting.

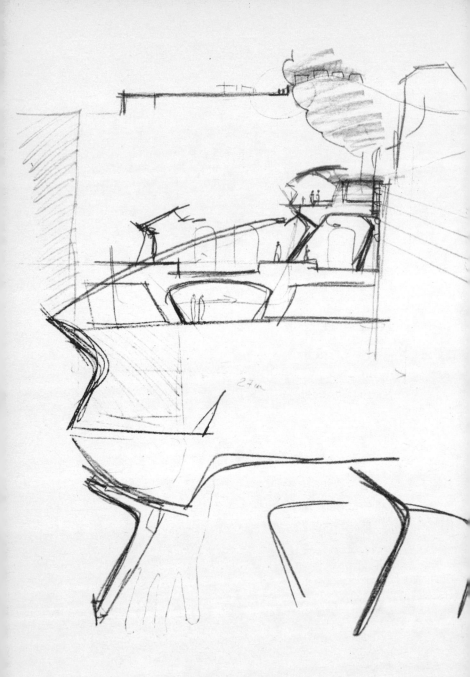

There were many other issues technical in nature that conditioned the construction process. For instance, the site on which we worked was very narrow—sometimes only twenty-seven meters, or ninety feet, in width. There were houses on both sides, so for technical reasons we had to go underground as much as fourteen meters, or about forty-five feet. Also, we had to build the project with the trains circulating the whole time; the traffic in the station was never interrupted, with a train arriving sometimes every two minutes. This was quite a challenge on such a tight site, especially in terms of staging and issues of safety.

I was very nervous because it was one of the first jobs in which I was responsible for the architecture and the engineering. This was a big thing, but I thought that we should also have some fun with the station. So in addition to relating the station to the functioning of the complex and to the urban context, I began for the first time to experiment with ideas of the body and of anatomy. I thought about gesture. I started with my hand and the idea of the open hand, which signifies sincerity and openness. From the open hand turned palm-side-down, I chose the area between the thumb and index finger as the shape of the column, which you then see repeated several times throughout the project.

I should also explain that the station is geodetic; it is not flat. The trains come out of a tunnel underground and go back into a tunnel. The two sides of the tunnel are curved downward in section, and only the center of the station is level. So we built the station slightly arched. The columns are inclined to remain perpendicular to the contour of the arch, but you don't see them as changing inclination one to the other. If they were vertical, you would immediately see it, but they are inclined in all directions. Even the underground gallery is like this. The station is also curved in plan. This was purely an engineering circumstance; the plan and section of the tracks were existing and had been done by engineers. But having a curve creates the possibility of making a station 280 meters long—approximately 900 feet—seem much longer than it is, because from the center platforms you do not see its ends. So the curves in both plan and section help the concept. You see, the shape, turning around like a torus, emphasizes the dynamics of the trains. Having this kind of rotational geometry makes it possible to achieve a dynamic effect in the architecture, especially through the repetition of elements.

It is interesting to observe that, in fact, the station is almost a non-station, because it disappears, hidden behind the roofs in the middle of the city.

I would like to make a small parenthesis to explain to you a different experience with concrete. I was charged by the Swiss Association of Concrete Makers to do a pavilion for an exhibit in Basel. For this project, I produced a machine that attempted to put concrete in movement. We know light concrete, we know heavy concrete, we know of all kinds of concrete, but concrete in motion is a rare thing. I don't know if you know Eugène Freyssinet, who was the inventor of the pre-stressing of micro-concrete to make wings for airplanes. This probably was never achieved, but he did all kinds of studies for this. This is to say that in the early twentieth century—in the '20s and '30s—people were extremely daring in their ideas and in their hopes for the use of this material.

In this pavilion I wanted to achieve forms that were extremely related to the body. By body, I mean the anatomy of the female and male bodies. I was interested in the idea of the sensuality of form. So you will see, maybe, ribs and circular elements that turn. It is a very tactile and soft way to use concrete that one does not usually consider. It is not only about moving the concrete but also about giving it sensual properties, like the properties of flesh or skin. The idea of "organicity" is even carried out in the details of how one piece joins to another. The shape of the detail becomes more than the result of an elaborate structural geometry; it permits us to recreate, as in a dream, the idea of the breast, for example.

The last project I would like to talk about is the Satolas Airport Railway Station in Lyon. You have to understand that this object was built by extremely careful and very gifted people using their hands to make the concrete. You have to understand the beauty of the construction site as a process, the way things are put together, the importance of getting close to the people working on the site, because they are fundamental. In fact, if someone risks their life it is certainly not the architect; it is the people on the site who do that.

The station is built in concrete, aluminum, steel, and glass, with granite paving. The concrete is white, using only white cement, natural aggregate that is taken from the area, and sand that is light gray. This gives a very light concrete, the color of which reminds you of the site. The roof is aluminum and reflects the sun. The façade of the train galleries is done using just pure concrete, with glass in between. The esplanade is made of granite cobblestones, and the handrails are all done in concrete. The large, open façades of the wing are in steel and glass. There are many different materials, but there are very precise rules for the way they are used. For example, the steel never touches the ground—it is always bordered by concrete. There is never a direct connection between a supporting element and the ground.

The wing of the station is oriented toward the south. We set the angle of the roof by the inclination of the sun on the solstice so that between the middle of June and November the sun does not enter the interior of the space. After November the façade lets light into the whole volume of the interior space. This permits us to have a building with very little need for extra climatization.

ando

centeno, plácitos vuestro

... instrumento físico organo
... la consideración ... los dados
... estima del espíritu se...

...peración fue a la puerta de acceso
mundo de las ideas cuya experien-
...ita e inmediata a el gesto de

ojo... habla
este
ojo... a
inter... as desajusques
orden de la intuición el orden del pensa-
...ecten ... otras imágenes juegan y con-
...ellas... la y prendida ... el ...

das   ←———————→ externo

I was interested in the idea of creating a structure that is based on certain proportions of the human body. Many of these bodies together support the train galleries of the station. The whole gallery is based on this modular.

The idea for the shape of the station and its roof was generated from the idea of the eye, and this is very important. Someone once said that if the painter Raphael had not had arms, he would have been a very good architect, because the working instrument of an architect is the eye. It is the faculty for seeing and judging and inventing things. There are two sets of eyes: the two eyes in front of us and the eyes of the mind that invent and combine things.

FORCE AND FORM

If we consider engineering an art—as I believe it is—and if we go back to a time when there was no difference between the art of architecture and the art of engineering, as I suggested we do at the beginning of my first talk, then we can consider that it is in ourselves, and especially in the new generation, that a rebirth of art happens. It is not only our heritage but also the mother of heritage that we must translate into action through our capacity to make buildings, reinventing them each time.

I would like to begin my second talk by showing another small object that is done using my children's toys. This one expresses another way of putting things together, hanging the same stone as in the previous case, but in this case, the forces are working parallel to each other; they do not cross. So it is like a game in which you discover that the weight of an object has the capacity to express something and that it all depends upon the ordering of the forces.

In my opinion, the consideration of the static properties of a bridge is very important. The bridge's static equilibrium is probably its most essential part, so that if we think of a bridge like a body, this static condition is analogous to the heart. Of course, the body has other parts; in a bridge you have other properties. The central part of the whole problem is certainly very much about resolving the problem of how to bring the forces from one shore to another.

One of the earliest bridges I designed was my diploma project done at the ETH in Zurich. The inspiration for this bridge comes from two different ideas. The first idea is of a cantilever bridge and the second is one of those beautiful arch bridges in which the forces are expressed by the arch. In a cantilever bridge, the forces work in another way. You have tension through the top and compression through the bottom, and then part of the bending moments are transferred into the shaft of the support. Beginning with the section of the cantilever bridge, I divided the material at the support so as to separate out these three forces, and I found that the idea of leaving a hole under the deck could be quite interesting.

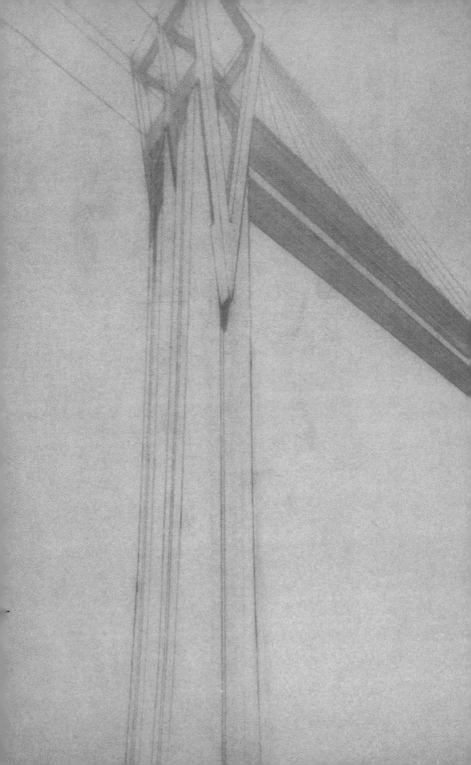

For the vertical support, I started with the idea of a pillar that has a certain proportion that today I relate to the human body. In fact, if you look at the pillar from the front it seems like the figure of a body with arms reaching up over his head to hold the deck. The lower support is located where this figure's head might be. Also, it is maybe a contradiction, but instead of making the bottom of the cantilever wider at the vertical support, I made it narrower to emphasize the independence between the upper deck, which needs to stay wide to conserve the dimension of the roadway, and the lower part of the cantilever, which is working in compression.

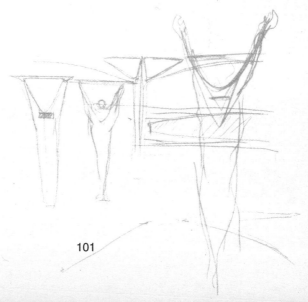

The first bridge I built and would like to talk about is the Alamillo Bridge in Seville. For the World Exhibition of 1992, my thought was to do two bridges that were symmetrical; one on one side of the island and another on the other side, with a viaduct linking them. First, I created the viaduct across the island—one that has many supports and is quite transparent. Then I generated the bridges in response to the scale of the space. These bridges were to have masts, which are the gestures that articulate this response.

For various reasons I could only build one of the two bridges. The design of the bridge was original; as far as I know, this kind of bridge had never been built before. Usually, in a cable-stayed bridge you have a compensation of the forces from the cables on the bridge side of the pylon to the cables on the anchored side of the pylon. However, if you incline the pylon, the forces are not only compensated for by the cables behind but also by the weight of the pylon itself. If the pylon is inclined enough and heavy enough, you can almost compensate the forces of the whole bridge purely through the pylon itself. This is what I tried to do here.

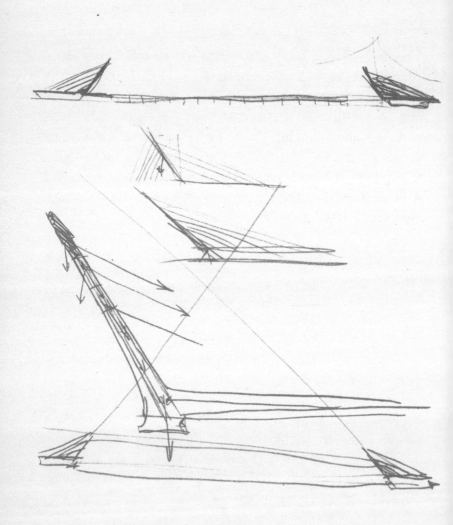

It is important to understand this object mechanically, and this will be one of the few times I will talk about this subject. You see, mechanically, what is happening is that each segment of the pylon or mast has a certain weight that pulls downward. Together with the tension in the cable, the resultant of these two coincides in direction with the mast. The next element of the pylon has then another weight—another force—and so on, and in each case the resultant stays in the pylon. Finally, the resultant compensates with the horizontal force coming from the deck and arrives as a pure vertical force into the foundation. The foundation is very small relative to the huge span of the bridge. It is interesting that, if the dead load produces a resultant at this point, then the live load will move it away. And then we have other problems, like the wind and many other things that we have to take into account in a cantilevering system.

Now I would like to return to my home town of Valencia and its river, the Turia. Valencia has a patrimony of bridges built in stone that, in my opinion, few other cities in this part of the world can claim; early Gothic, high Gothic, Renaissance bridges, all done using arches. These are magnificent works that we consider, without a doubt, as having architectural value. In fact, if you look at the old stone gates of the city and the Gothic bridge that leads to these gates, you see that there is very little difference between the architectural language of the gate towers and that of the bridge. There is another bridge that has not only huge public stairs but also chapels, and another—a concrete bridge from the early twentieth century—that has caryatids and other decoration executed by Terencio, a sculptor from the city. This is to say that if you go back a little, you come to a time when architecture and bridge building were absolutely linked together. The bridge builders were very conscious of this particular relationship. They were also conscious of the seriousness of the act of constructing a bridge. Many of those bridges are five hundred, six hundred years old and have resisted unbelievable floods. But even while taking into account the engineering needs of the bridge, the builders also thought about making small chapels (maybe to pray in, so that they would not be swept away in the next flood).

In this context I built a contemporary arch bridge over the Turia. It has a plaza below it and an underground station below the plaza. It is a pure steel bridge as much as the others were pure stone bridges. (I also built another bridge for Valencia, just outside of the city. Except for a few elements, the whole bridge is done in concrete.) In this one I divided the bridge's ramp into two, because first, it had to be very wide for traffic, and second, because it was the continuation of a boulevard that comes onto the bridge from one side. I chose to leave a void down the center of the bridge that corresponds to the space of the boulevard. In the space underneath the bridge, I wanted to emphasize the spatial effect. A pool of water is to be built in this lower space that will reflect light onto the underside of the bridge and also reflect the bridge itself. Light comes into this space from the sides and from above, between the pedestrian deck and the roadway.

Next is the Felipe II Bridge, built for the city of Barcelona. I would like to emphasize in particular with this bridge the capacity of a public work such as a bridge to generate infrastructure and in doing so, change the circumstances of a part of the city. The Bach de Roda area where this bridge was built was a very poor neighborhood—a kind of *bidonville*. The people who were living there were moved into better apartments so that the area immediately adjacent to the bridge and the train station below it could be turned into a series of parks. So, in a very rough part of Barcelona you now have a place that has the potential, in my opinion, to become a significant part of the city. In this case, the need to build a bridge and produce a link provided the impetus for regenerating a portion of the city.

One of the ideas of this bridge was to make a place. I made the bridge wider in the center, like a balcony, and the arches that span the railway tracks signify not only the bridge itself but also this place in the middle of this neighborhood. In this part of the city the landscape is not very romantic, but on the other hand, when I look at the mountains behind and the housing around, this space has huge potential to become a place of interest in Barcelona, especially now that there are green areas on both sides of the bridge.

I think that all cities need very much their public works to help them regenerate, especially now. During the '70s, architectural and urban interests focused on the historical point of view, which supported the regeneration of city centers to make them more habitable. Those interests also initiated the preservation of a lot of buildings that otherwise would have fallen victim to speculative development. But meanwhile, the development of the city's periphery began to become a very significant problem. So today, the problem is how to reform those parts of the city—and particularly European cities—where the population has doubled or tripled during the last thirty or forty years. I think that public works like bridges and stations can become very significant forces in regenerating areas by creating and focusing urban activity.

These next projects explore the idea of the arch. The first is a trade exhibition hall in Tenerife, in the Canary Islands. Sometimes when you design a functional building that, like many of the bridges, has to be done for a very low cost, the idea of using the arch is important. It is very efficient for large spans. In this building there are many different types of arches: concrete half-arches, and, on the top, steel arches. A large arch spans 240 meters—something like 800 feet—to support the whole roof.

I would like to talk about the inclined arch in particular. The first example I will now show is the La Devesa Pedestrian Bridge in the city of Ripoll, which is in the northern part of Catalonia in the very dramatic landscape of the Spanish Pyranees. In this landscape we built a bridge to link the train station to a residential neighborhood located on the other side of the river Ter. The area between the bridge and the edge of the neighborhood was also transformed into a park, with a plaza and a series of interventions that were built by other architects. As I said earlier, bridges are very powerful when you want to regenerate a place, because they introduce a very good reason to restructure the surrounding area and, in so doing, make more livable these parts of the city that are rather lost, like this one here in Ripoll.

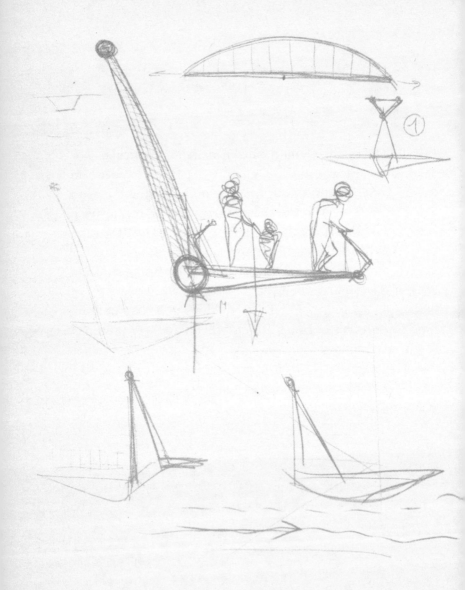

I would like to focus on the cross section of the bridge and the inclined arch, because there is a progression toward a rather bold asymmetry in the arched bridges as I have attempted to exploit the phenomenon of torsion. In the bridge in Ripoll, because the support—the arch—is offset to one side of the bridge, the weight of the bridge and its load create a torsional moment about the point of support. This torsional moment is put into a pipe section that runs the length of the bridge and connects the ribs of the deck to the inclined arch. In the bridge in Valencia, we transformed this pipe into the whole deck, so it is the deck that resists the torsion. In the case of Ondárroa, we have a lateral box girder and a cantilevering pedestrian walkway from which ribs spring to hold the arch against buckling, and from the arch we have tensional members that hold the lateral box. In the case of Orléans, the torsion is taken completely by the section of the roadway, which the arch supports just by cables.

So there is a kind of progression beginning with Mérida, where the arch is centered above the roadway deck, which is the element that provides torsional resistance. This is a large arch with three cords and is case number one. Case number two is Ripoll, which was for me a kind of experiment to control the system of the inclined arch in a span of 70 meters, or 230 feet, making it feasible at a very low cost and with a deck that is only something like 3 meters, or 10 feet wide. Case number three is Ondárroa, in which there is a significant traffic load, and case number four is Valencia, with four lanes of traffic. Orléans is case number five, with a major span, four lanes of traffic, and pedestrians on both sides.

What is interesting in this type of bridge is the torsion. In many of the standard cross sections that we see in arched bridges, the arch is straight up and so the torsional stiffness that we have in the box girder that supports the roadway is mostly unused, because you have only the unilateral load. What I have tried to explore in these bridges is the phenomenon of torsion—how to exploit the torsional resistance of the roadway to create a certain asymmetry in the bridge that permits me, for example, to emphasize the position of the bridge in relationship to the city around it, or the direction of the water, or even the position of the sun. It permits me to sensitize the bridge itself, as a phenomenon set into the surrounding landscape.

The landscape of Ondárroa is very picturesque with the Atlantic ocean in front, a small harbor, and fishing boats. For the Puerto Bridge, I tried to take advantage of local conditions and materials, using the stones of the area to build the embankments, for example. Many people walk across this bridge to Ondárroa's only beach, so it was necessary to provide an ample walkway. This brought me to the idea of making a big balcony on one side of the bridge. There is another pedestrian walkway on the other side of the bridge, but this one is given major importance by separating it from the roadway. With the hangers of the walkway, I made the bracing for the arch, which supports the bridge with cables. As with the bridge in Valencia, I was also interested in the space underneath the bridge. There are very large changes in the tides in Ondárroa and many people come with their boats to the area that surrounds the bridge. It was necessary to build big staircases going down into the water as part of the embankments. I separated the parts of the bridge so that the light can enter the space underneath, and this, together with the embankments, makes a very interesting space.

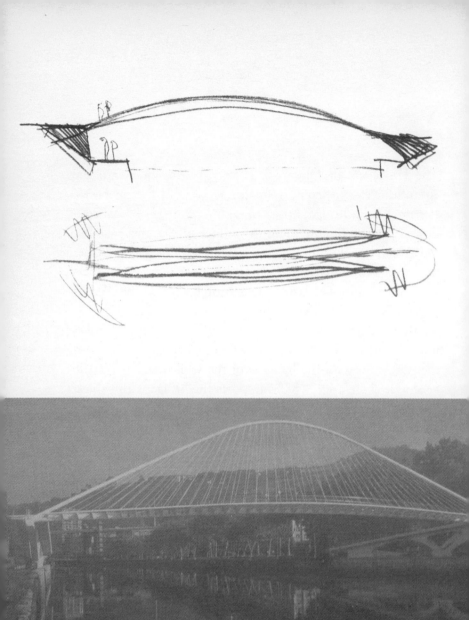

The Campo Volantín Bridge that I built in Bilbao is interesting because it is somewhat the antithesis of what I had learned—or thought—a classical bridge was, especially in terms of the embankments. If you think about a classical bridge, like the bridges in Venice, you have an arch and then all the forces of the arch are brought directly into the ground at the embankment, which is solid. This event is signified with a staircase, which permits the people to descend to the canal. The embankment is a classical element of the bridge; it is the way the bridge touches the ground as a continuation of the forces.

In Bilbao, I supported the bridge on cantilevered sections that rise up from the bank of the river and run parallel to it on both sides. The bridge deck is placed on these supports, which are like arms. The cantilevered section of the support is a half-arch, and so, where you usually would have had solid embankments, there are now voids. These voids and supports give a directionality to the bridge—one that can be associated with the direction of the flow of the river.

The flow of pedestrians runs parallel to the river, crosses the river, and then returns in the original direction. I signified this movement in a very simple way, by making a curve in the plan of the bridge. So the idea of the strict, straight line of a classical bridge disappeared as well. I wanted to make an element to hold this structure that—although asymmetrically placed—would optically compensate for the asymmetry of the deck. So the arch inclines over the deck counter to the curve.

The torsional pipe under the deck runs straight, so the particular point in which the arch and the torsional pipe come together is very important. Because the force of the arch is coming into this point diagonally—in section—and the force of the torsional pipe is coming in horizontally, the resultant of the forces displaces the vertical support necessary by several feet, revealing, in a way, one of the many paradoxical situations in the bridge: the support for the arch is three or four feet away from where it joins the torsional pipe, although it is centered in terms of the forces. I tried to play formally with a very pure understanding of the way forces of construction work in this bridge. For example, if you look at the plan of the bridge, you will see that the torsion is compensated globally, because the deck on one side of the torsional pipe is equal to the deck on the other side in surface area. At the end, we have the same torsional forces side to side.

Today we have somewhat lost the idealism of the '60s, when architecture was very much devoted to social problems. In fact, today we ignore these problems, yet we live in a world in which a third of the people do not get enough to eat every day. I mean, it is quite dramatic. Can you imagine how much infrastructure is still needed? If you think back to the heroic times of engineering, when people arrived at new places, imagine what it took just to bring water to a place, or to stop a flooding river, or to create sanitation infrastructure. I think you still feel the strength of this need when you build bridges, especially bridges in cities.

I also think that the potential of bridges and bridge design has not yet been achieved. The vitality of bridges comes both from necessity and from the fact that they are unbelievably significant, unignorable elements of the city. Could you imagine New York, for example, without those magnificent bridges? What if the George Washington Bridge was a multi-span bridge, instead of being this gesture jumping over 1,000 meters—more than half a mile—when the previous span had been just half that, or even less? Engineering can still provoke very wild and strong responses from gestures like these.

If you look at engineering design of the nineteenth century, you see that engineers took a lot of care in designing the handrails and the lighting. They were conscious that to take care of the details was to give more to the bridge. There are people that say "architecture is all that you can take away from a bridge to leave the bridge standing." This is not true. Architecture is the bridge itself, because the bridge is dedicated to man. All that gives satisfaction to man is good for a bridge.

MOVEMENT AND FORM

This evening I am faced with the task of giving you the third lecture, which is probably for me the most difficult one, because I would like to convey to you, in a way, the most intimate part of the work. This part is the cumulative result of ideas, sculptures, and drawings that have generated this or that building. In showing you these, I am acting somewhat like a cook who wants to give to other people—especially to young people—his secrets; what ingredients, what kind of herb or spices he has been mixing in to give this or that particular taste. Although the work is very personal in nature, it is also the culmination of many things that I have learned during my life...the approach of different artists...the approach of the idea of art...of the art of engineering...of the art of architecture, and of how these things can be linked together. So what I want to give you is quite a lot. For this reason, I will speak with very simple words, because that is the easiest way to communicate most directly.

In the first two lectures I began by showing small sculptures made from my children's blocks. The sculptures were not very big—something like one foot wide. In both, the problem was how to hold a stone hanging from these machines. If you look, it is only the change in the placement of the bearing (I want specifically to use this word, which is an engineering word) that creates two different ways of ordering the forces. In the first case, the forces are working parallel, and in the second case, they are crossing each other, but, in fact, it is exactly the same exercise—how to hold a stone. Relative to this, there are several things that I would like to emphasize. First, to hold a stone is, itself, something important, because you are working against the tendency of the stone to fall. Like the scientist who allegedly was hit on the head with the apple, we live in a field of forces and gravity that is immediate to us. Probably our bodies would be built completely different if gravity, or the forces around us, were other than they are.

Second, the sculptures hold together because the weight of the stone is activating the system. This means that, if the stone was not there, it would be very difficult to hold the machine together. So the weight, or the permanency of gravity, is also something that is materially present. The sculpture stays together because gravity is there and because gravity is a constant situation. What would happen if I were to cut the cord that transfers the weight of the stone through the whole system? It is very simple; the stone would fall down, and then, also, the small structure. When the stone falls down, a movement is produced. The forces of the machine are very much related, in their static condition, to this movement. This means that the presence of force is always related to a movement that cannot be established prior to its activation. Consider for a moment that forces are like crystallized movement. This is, in my opinion, quite a beautiful understanding, because even in its static condition, in the most stable thing, movement is hidden. A movement is there, and because of that I went from the idea of force and form, which I talked about in relation to my bridges in the previous lecture, to the idea of movement and form, which is much more difficult to represent in slides.

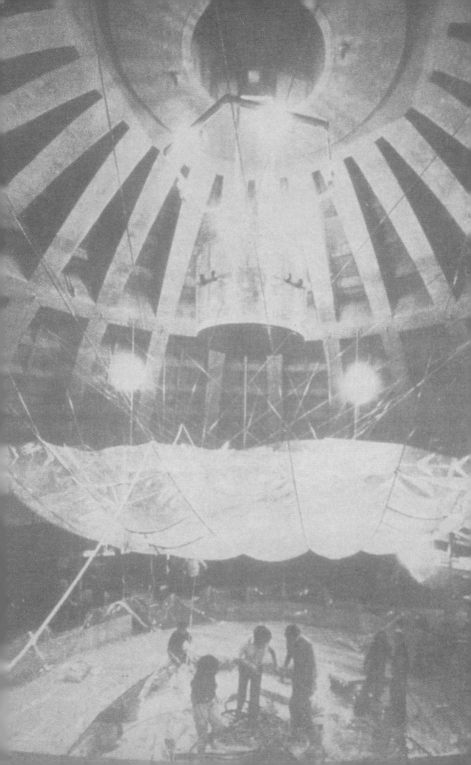

I would like to start by talking about two projects that were done while I was a student at the ETH in Zurich. The first project was done with a group of students under the direction of Jürg Altherr. We built a swimming pool hanging from the cupola of the ETH. It was hung by twenty-four wires that supported twenty-four ribs and a skin of polycarbonate that was 1.2 millimeters thick. It contained twenty-four cubic meters of water, and it was even possible to swim in it, although it was not possible to remain in the water for long because the pressure of one's feet might have broken or deformed the membrane. The construction was a very particular challenge, because below the cupola, under this swimming pool, was the library. Can you imagine? All that water over the library?

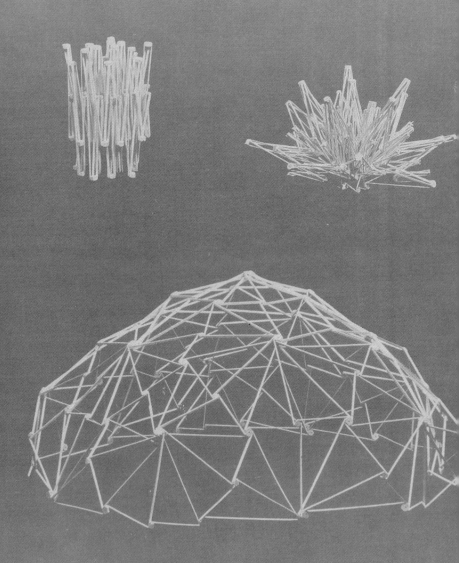

Different from, but not unrelated to, this idea of holding something heavy, the second project was a study of the explicit movement of polyhedra. My doctoral thesis was called "On the Foldability of Frames," and its basic focus was on the study of topology and the way in which a polyhedron—a very complex polyhedron—could be folded or transformed to become a bundle in which all the lines are parallel. Through a series of phases, this bundle opens, slowly changing its shape to recuperate the major polyhedron, which approximates a dome. Even if it did not look extremely good, the focus was on the complex geometrical process of changing shape from bundle to half-sphere.

Another source of interest that I would like to emphasize is nature and the very simple observation of nature, meaning to look in a straightforward and pure way at the natural objects that are around us: trees, grass, flowers, or whatever the natural object. From a group of study sketches, one early model that I made attempted to take the structure of a tree and apply it to an idea for a group of trees using steel and glass. I was no longer thinking in terms of the specific qualities of the object itself, but rather, through a kind of systematic approach to the object, I decided to make them transformable, meaning in this case that they change their form. So the ribs at the tops of the trees—configured as the roof—hinge at the point at which they are supported, and the entire roof opens through a mechanical transformation. The idea of the trees was later incorporated into a project for a gallery in Toronto. The idea of mechanical transformability, however, was not incorporated into the trees but into a very large, operable window that separates the gallery, which is quiet, from Heritage Square, which is a noisy, animated space.

Continuing with the idea of the tree—an idea that is quite general, in my opinion—we built the Orient Railway Station in Lisbon. Here, I used the same vocabulary, specifically choosing it because the city of Lisbon is not only very beautiful, but also has the very, very soft character of cities on the Atlantic coast. The space is very transparent, with a beautiful quality of light. In this context, I did not want to do a very strong structure, like in Lyon, but a soft one—an open structure with the trains passing on one side and the view passing on the other side.

Another topic that is also very important in architecture is anatomy and the idea of reading in the human body structures, or appreciating in the human body a sense of architecture. Whatever we do, the magnitude or the dimension of a thing is always related to our bodies. Architecture, in a very natural way, is purely related to humans, because it is done for—and by—people. This makes anatomy a very powerful source of inspiration. And this was true not just in the Renaissance, when the human anatomy was the basis for rules and proportional systems, but also in the twentieth century with the Modulor. Anatomy—the idea of the hand, of the open hand, the idea of the eye, the mouth, the skeleton—is a rich source of ideas and inspiration. In the tectonics of our own bodies, you can discover an internal logic that can be valuable in the making of buildings.

It is important to recognize in the phenomenon of architecture its purely plastic or sculptural aspect. This is not in conflict with the functional aspects of architecture, nor with the structural aspects. The wings of the station in Lyon, for example, take their geometry from a sculpture that was done previously as a study on the eye.

More abstract than my studies on the eye are my studies on how the head is supported in a position over the shoulders. Why and how can I turn my head? The entire mass is supported only by the atlas of the vertebral column, and so the head can move. This wonder of moving the head—rotating it, inclining it, or both simultaneously—is quite interesting and, in a way, quite dramatic. I have been studying how to hold the head. I begin with a pure volume and mass—a cube—trying to hold it with a minimum amount of elements; for example, with a very, very thin spindle and a series of cables around it. In another sculpture, the mass, or head, is supported vertically by one element and a second oblique element is used to push it back—to fix it in its horizontal position. In the section of the Stadelhofen Station, the same principle is used, but the mass that is represented is a cube and the sculpture is the earth.

I like very much the purity of a single idea, just as the pure expression of a single note can be a very powerful thing. Complexity comes from the superposition of ideas in a coherent way. This means that, although each one of those ideas is capable of living independently, you can also put them one over the other, like a painter who works not only with black but with several colors, or hides a lot of blue behind the painting in order to capture the horizon. So, for example, in Stadelhofen, in addition to the reference to the

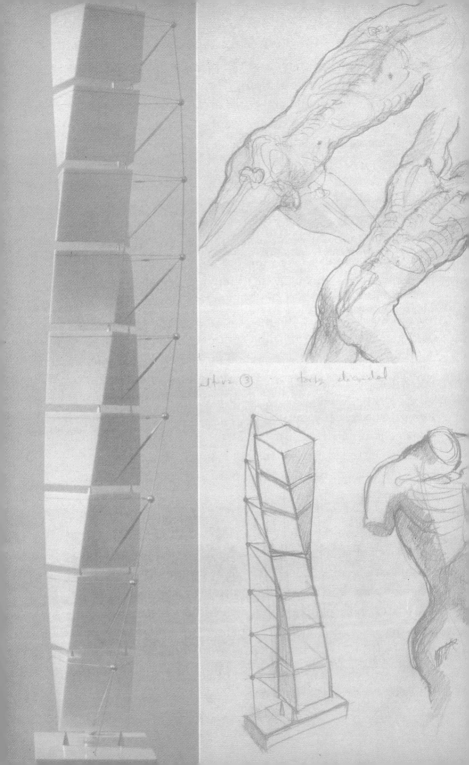

propped head, there is also the idea of the hand—the open hand. This reference is represented throughout the project. It became a leitmotif for the configuration of most of the structural members of the station: the main buttress supports, a small canopy, the pergola. The hand mirrored makes the cross section of the underground. In many places there is a gesture that can be associated with this same geometry.

Another important series of plastic studies goes a step further. They describe the spine, or how our body stands up. The spine is made up of vertebrae that are represented in the sculptures in a very elemental way, as a series of cubes.

Once you have the concept of the spine, it is very easy to move the elements around, reshaping the spine in several ways. In one case I staggered the position of the cubes around the central cord. In another they are stepped out so that the idea of movement is much more explicit. Although the material is the same, the hieratic rigidity has disappeared, so that you now have more the idea of a curving spine. Also quite important is how our spine twists, how it turns around an axis, and how it bends and reaches. What before was quite a shy movement becomes very explicit.

In some later studies, I changed the geometry of the cube in order to make the sculptures bigger. In one project that is twelve meters, or forty feet, high, I wanted to create a more organic shape, and so I moved away from the cube, using double pyramids, or octahedrons, instead. There is a series of seven of these octahedrons in concrete, one after the other, each held back by two pins and a single cable. The seven pieces are counterbalanced by two very long legs. At this scale, the sculpture is now more than an abstract form; it is also a construction principle. This sculpture became the basis for the bridge I built in Seville.

Starting with a very pure and abstract study of the human body and anatomical relationships of weight, one can move through to a major problem in steel. When you start building a sculpture that is twelve meters high, the construction problem begins there. How to hold things, how to choose the material; these issues become critical. You have to deal with concrete. You have to deal with cables. You have to deal with supports and with tensions, with transport, and with many other things.

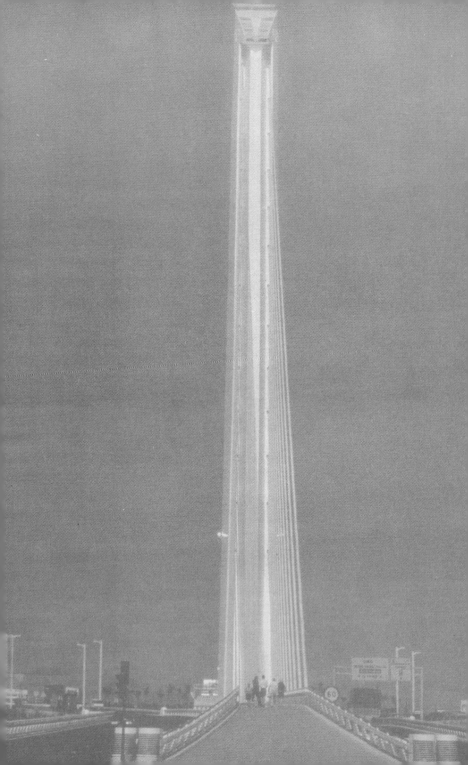

In particular views of the bridge in Seville, you can recognize the very abstract origin of the bridge's idea. When you look at the bridge from the front or from the back, you do not think of a bridge. Instead, it is the expression of something autonomous. A plastic or sculptural exercise is very beautiful in itself, because you feel free. You are only bound in terms of the limits you set, saying, for example, "I would like to be extremely orthodox. I will work only with cubes." Whatever the specific terms, you have limited your vocabulary, but you are still free, because the only goal that you are pursuing is a pure plastic achievement. When you are dealing with a bridge or a building, even if it is a plastic event, you are bound by functional needs. On the other hand, you have a big advantage, which is scale. No sculpture in itself will ever reach the scale of a bridge or a building. It is this that gives architecture— and particularly, architecture that is integrated with engineering—its significance.

Here I would like to talk about some research that is related to movement, but more specifically, to simple structures that can move. We built a pavilion for the State of Kuwait at the World's Fair in Seville, and it is done using a series of wooden half-arches that can open. This structure covers a terrace, the floor of which is made of translucent marble. Below this terrace is the interior part of the pavilion, and during the day sunlight is filtered through the marble to light up the interior space. The roof elements are supported by concrete members, and each one of them can be activated by an individual motor so that very slowly the whole roof opens and transforms. Because the elements are independent, you can control the different stages of the roof's opening. In the movement of the roof, one can imagine the hands with their fingers bent down, serving to protect the space, and then opening, palms up, to cup the sky.

A sculpture that we did for the courtyard of the Museum of Modern Art in New York was placed beside a weeping willow. The branches of the willow are gently curved. The branches of the sculpture are also curved, and they fall slowly—each one turned by a wheel—to almost touch the sculpture of *The River* by Aristide Maillol, which is below, on the surface of a pool.

From a series of sculptural studies that were based on the idea of movement—and specifically, the idea of topological surfaces that are generated by straight lines that revolve around multiple centers—came another set of studies for a possible roof structure. In these sculptures, straight lines generate folded, curving surfaces. In subsequent architectural studies, these straight lines became single elements of construction. You have a shape on the ground, then one central inclined ridge line, and the identical construction elements connect the profile of the shape—which is a circle in one case and a half-ellipse in another—to the inclined line. This same study was also a source for ideas about how to make this type of form move and open. The two halves connected by the central line are like two hands that are hinged together at the thumbs. They open up and close down around the axis of this hinge.

The extension for the Milwaukee Art Museum, which we are now building, uses a variation of this roof structure. In my project, I linked the original building by Eero Saarinen and the extension by David Kahler, built something like twenty years ago, back to the city with a bridge. This existing museum is, itself, like a bridge, and, in my opinion, what I have done is very respectful of the idea of this museum and its relationship to the city. In the existing museum you have a bridge and a sculpture-like volume in front of the lake; in my project you also have a bridge and another volume in front of the lake. While the existing volume is compact and closed, this new volume is transparent. A very shallow shed building links the new extension to the old extension, permitting from the height of the bridge a view of the lake's horizon beyond.

So as to conclude with architecture, I will speak about the cathedral of Saint John the Divine, which was never completed. It was started by Heins & Lafarge and continued by Ralph Adams Cram, with some work done on the vaults by Rafael Guastavino. Only the nave and apses exist today. In a competition entry for the cathedral, I tried to implement some of the vocabulary you have seen before but in a more symbolic way, because symbolic language in a cathedral is very manifest. I thought of comparing the cathedral to a tree, with the roots at the bottom, the trunk, and then the foliage at the top. One task of the competition was to create what they called a "bio-shelter." The bio-shelter was supposed to be on the interior of the cathedral, but I thought it better to put it above. The vaulted space in between the cathedral's interior space and the roof is usually dark and closed. I wanted to open it up.

The idea I proposed was to replace the roof, which is temporary today, with a glazed roof and to plant trees in this upper space so as to create a garden over the cathedral. The garden would be a reproduction of the temple itself. It is interesting that in the work of Beethoven—for example, in the fantasy for choir, orchestra, and piano—they sing about *Im Tempel der Natur*. Nature is considered a temple. And so we were thinking about making this temple of nature. This is a very romantic idea.

The plan of the cathedral follows a Latin cross. One can look at the human body as a temple, and inside the shape of the Latin cross can be found the idea of the human body. So the garden and the body are superimposed and related to the geometry of the cross. This kind of mysticism that comes from a far away time was part of the conception of the building.

I wanted to make the glass roof of the garden operable, so as to let the rain water collect in the garden. The triangular sections of the glass roof function in a way that is similar to the big windows in the gallery in Toronto. They just turn around an axis. The roof was designed with a very high spire, which could be used for thermal purposes, to create a microclimate on the interior of the roof.

To approach the natural world with respect, to approach the landscape with much more respect; this is a concern of architects and engineers. And I would like very much to underline this, without going too far into this theme, but the integration of buildings into the landscape is something very important. I think that in the basic idea that we submitted for Saint John the Divine there was an interesting equilibrium.

CONCLUSION

All these projects in which I have been working have been possible not only because of the people who have executed them—and who I appreciate very much—but also because many people have been helping with their hands and their eyes. Making models, making drawings, in my office. They have all participated in this effort and have contributed a lot to the projects. My wife has helped me very much in these talks. Thank you very much to everybody.

This defines the end of my talk. I would like to thank you very much for your attention.

SANTIAGO CALATRAVA VALLS is one of the most important figures working at the intersection of architecture and engineering today. Born in Spain, he maintains practices in Zurich and Paris. His innovative and undisputedly expressive bridges, train stations, airport terminals, concert halls, and art museums—and the remarkable sketches he creates while designing them—have been the subject of numerous books, including *Calatrava, Public Buildings* and the two-volume *Santiago Calatrava's Creative Process*, both published by Birkhäuser.

CECILIA LEWIS KAUSEL is professor and director of the Interior Design Department at the Chamberlayne School of Design at Mount Ida College in Newton, Massachusetts. She received BA degrees in biology and physical anthropology from UMASS and an SM from the Department of Architecture at MIT. A guest professor at the Bauhaus and a research affiliate at MIT's Department of Civil and Environmental Engineering, Lewis Kausel has published numerous journal articles and a report on the preservation of the Alhambra for CEDEX, in Madrid.

ANN PENDLETON-JULLIAN is an architect and associate professor at MIT. After studying astrophysics for three years, she received her BArch from Cornell University and her MArch. from Princeton University. She is the author of the award-winning book *The Road That Is Not a Road and the Open City, Ritoque, Chile* and several other theoretical works. Her architectural work has been published and exhibited internationally.